THE
WATCHMAN'S
RATTLE

THE WATCHMAN'S RATTLE

Thinking Our Way Out of Extinction

Rebecca Costa

Virgin BOOKS

Published in 2011 by Virgin Books, an imprint of Ebury Publishing
A Random House Group Company

2 4 6 8 10 9 7 5 3 1

Copyright © Rebecca Costa 2010

This edition is published by arrangement with Vanguard Press
a member of the Perseus Book Group

www.virginbooks.com
www.rbooks.co.uk

Address for companies within The Random House Group Limited can be found at:

www.randomhouse.co.uk/offices.htm

The Random House Group Limited Reg. No. 954009

A CIP catalogue record for this book is available from the British Library

ISBN 9780753539781

Penguin Random House is committed to a sustainable future for
our business, our readers and our planet. This book is made from
Forest Stewardship Council® certified paper.

Printed and bound in Great Britain by Clays Ltd, St Ives plc

For Bailey, Ben, and Camden

Contents

Preface

by Sir Richard Branson

Since becoming a voracious reader of books about science, the environment and history over the last twenty years, I have always wondered why sophisticated societies collapse. It has been a fascinating riddle that historians have never explained well. The Marxists always saw it in revolutionary terms and the environmentalists have always seen it in climate terms, while the Malthusian biologists/historians have always seen it in population terms. Together they've explained the collapse of sophisticated cultures as a function of social pressures, population explosion and an adverse environmental kicker pushing a culture over the edge.

That view has always worried me and something sneaking in the back of my mind has often wondered, "Why did clever people who could build pyramids and massive cities in different parts of the world manage to allow that to be destroyed over a few generations?" If great ideas had built civilisations that advanced well beyond the flowering Tudor England of the 17th century, what other factor was involved?

With her book, *The Watchman's Rattle*, Rebecca Costa has made an extremely brave, spirited and well-informed attempt to answer this question in the context of the seemingly insoluble problems facing a global civilisation of 2011, with its 6.5 billion human participants who are currently having enough babies to guarantee that it will be a 9 billion person problem by 2040.

It is one of those rare books that one picks up and knows within the first few pages that it is extremely important, but more than that it is also a fascinating read that challenges our very beliefs about how we solve the problems we face in a resource-constrained world with a rapidly growing population.

Even in my own experience of modern business, the most frustrating thing is people telling you why things can't be done based upon beliefs rather than data, reaction to the past rather than analysis and a simple view that it is easier to file things in the "too difficult cupboard" rather than trying to manage the risk of achieving them. I remember when we were trying to get Virgin Trains off the ground in the UK with a new technology tilting train that could go much faster round the bends of the old UK rail network. All the experts said it couldn't be done and the civil servants hated the risk. One almost got the impression that people thought it would be better to do nothing and pass the problem on to the next government. How was it that the country that had invented the railway and the first high-speed trains could have such entrenched attitudes?

I've always believed that we will not get to where we need to be with short-term fixes that pass one issue after another on to the next generation, with each successive one left with less room for manoeuvre. Energy and its clean production is the key to a continuously developing civilization and Thorium fission reaction, clean fusion power, artificial photosynthesis or indeed solar power from space are all potentially viable solutions.

And if you want to know why none of them are being pursued with anything like the investment, vigour or attitude to risk needed, then this book will help you answer the question. It will also help you reason with the fact that at the same time as the above solutions are largely ignored, such enormous investment is poured into marginal and expensive hydrocarbon extraction despite the fact we now know we only have enough left to last 40 years tops!

Rebecca Costa has successfully challenged the prevailing view that our historic, scientific and economic progress since the dawn of capitalism will always be there to rescue the boat and she has done it by introducing our recent breakthroughs in our knowledge of neurological science to convincingly argue that we literally need to change and adapt the way we think about increasingly complex problems that the human ape has to solve in the next century.

I have taken what Rebecca has said very seriously and whether it be in space, in the air or on the ground, I for one will look for solutions in future rather than falling back on simple truisms that may not be true. Let's not leave it to future generations to blame us for a civilisation collapse we now have the tools to prevent, even if our ape-like brains don't want to admit it!

Foreword

by Edward O. Wilson

In *The Watchman's Rattle,* Rebecca Costa presents a view of the parlous human condition with which I completely agree. The clash of religions, and civilizations, she argues, is not the cause of our difficulties but a consequence of them. The same is true of the global water shortage, climate change, the decline of carbon-based energy, our cheerful destruction of the remaining natural environment, and all the other calamities close to or upon us. The primary cause of all threatening trends is the complexity of civilization itself, which cannot be understood and managed by the cognitive tools we have thus far chosen to use.

We have come to this point, Ms. Costa tells us, because humanity lacks an adequate sense of its own history. We have not faced honestly the central questions of philosophy and religion, which are scrawled in simplest terms on the canvas of Paul Gauguin's Tahitian masterpiece: Where did we come from? Who are we? Where are we going? By *history,* Ms. Costa correctly means not just of this country or that, but of the rise and fall of past civilizations and, beyond, the six million years of biological evolution of the human line, played out in intricate relationship with the rest of the biosphere.

From the long haul of biological evolution came genetic human nature. This period, during which we acquired our emotions and cognitive capacities, shrinks to an eyeblink the human history begun with the Neolithic revolution ten thousand years ago. The past three millennia have seen the exponentiation of everything gained by cultural evolution: population spread; the efficiency of work, knowledge, technology; and, unfortunately, the depletion of natural resources; the destruction of the remaining natural environment; and an increase in the war powers of more and more groups and nations.

Professional optimists like to say that doomsday predictions are as old as the written word and never come to pass. They believe that the genius and spirit of humanity have always found a way around its problems, and will again, and yet again. In short, not to worry. But to think this way is to ignore the reality of exponential change. If you have a doubling time of any entity or process of, say, twenty years, for a great many such periods the world as a whole will remain unsaturated and manageable. But at some point in any exponential growth, the next doubling time produces an absolute increase that overshoots all the space and resources left. At that point, the options for accommodation also shrink drastically.

There is great truth in the oft-quoted riddle of the lily pads. A pond (a lake, an ocean, all apply) starts with a single lily pad. Each pad doubles per day; the pond will be full in thirty days. When is the pond only half full? On the twenty-ninth day. After the next day, the thirtieth, further growth is so fast it will, if somehow continued, overwhelm the pond and everything in it in a matter of hours.

I am on the side of Rebecca Costa and others—let us accept the title of realists-in-search-of-a-solution, not doomsayers—who say that because of exponentiation, humanity doesn't have a lot of time to figure things out. We have to solve our problems not by continuing to use emotions and responses that suited our primitive ancestors but now put us all in imminent danger. Instead, we need to use knowledge and reason and take an honest look at ourselves as a species. We need to grasp the increasing complexity of our social and political arrangements, and reach solutions. In *The Watchman's Rattle*, Ms. Costa urges us to do so by employing the better instruments of our genetic nature.

Introduction

ONE COLD AND RAINY SPRING DAY, I was sitting in E. O. Wilson's office in the back of the Natural History Museum at Harvard when he turned and said, *"It's dangerous to state the obvious."*

It was a chilling warning from the most acclaimed naturalist in the world and the only scientist to be physically attacked on U.S. soil for his views.

Wilson was foreshadowing what lay ahead: the scrutiny, criticism, irrational opposition, and attempts to disparage me for using terms such as "evolution" and "biological obstacles" to explain why governments, leaders, and experts have become gridlocked.

But everyone knows we can't fix our problems anymore. The world we're handing off to our children is in much worse shape than the one we inherited. Something dangerous is happening, but we haven't been able to put our finger on exactly what it is.

So in his charming southern way, Wilson was simply confirming what I suspected all along: Once we discovered the reason for gridlock, it would likely look and feel *obvious*. The truth always does.

That said, it would have been impossible to write *The Watchman's Rattle* until 2006. Six pieces had to fall into place and they were 150 years in the making.

The first piece of the puzzle arrived in 1859 with the publication of Charles Darwin's *On the Origin of Species*. Darwin uncovered the slow, continuous pace at which all life-forms, including humans, respond to their environment to increase their opportunities for survival.

To this day, his discovery remains the most important scientific principle governing life on earth.

Then in 1953 came the discovery by James Watson and Francis Crick of the double helix in DNA. Together they unlocked the mechanics of *how* Darwin's theories worked, and for the first time it became possible to trace the biological genesis of all living organisms. Overnight, evolution graduated from a widely embraced principle to a provable fact.

By the time E. O. Wilson's controversial book *Sociobiology: The New Synthesis* came along in 1975, I was already a junior at the University of California. According to Wilson, genetic inheritance plays a big role in how we behave as individuals and in groups. Human beings are not born "blank slates." We are born with hardwired predispositions and instincts aimed at assuring our species' survival. Natural selection has a big hand in explaining modern aggression, altruism, hoarding, competition, even mate selection.

Then one year later, Richard Dawkins released his book *The Selfish Gene,* which took a harsher dog-eat-dog approach. According to Dawkins, individual gene pools, not the species, are fighting to survive. Genes *selfishly* want and need to perpetuate. To this end, they manipulate human behavior to assure continuity, even to the point of being destructive to the group.

The revelations of Wilson and Dawkins turned my college education upside down. Torn between my love for biology and a desire to find a humanitarian career, I spent months arguing with the university about changing my degree from sociology to sociobiology. But the university fought back, saying, "Two books, no matter how historic, don't justify a new degree." So in 1977 I, along with a handful of students, reluctantly accepted the first bachelor of science degree offered in "combined social sciences."

Following this hollow victory, I made my way home, where the fifth piece of the puzzle was unknowingly taking shape: *the accelerating complexity of the human condition.*

In the 1970s my family was living in Northern California, an area populated by apricot and plum orchards, unspoiled creeks, and hills covered with dry grasses and century-old oaks. No one suspected this rural landscape would soon become ground zero for a new era of technology. But by the time I arrived in Santa Clara in 1977, "Silicon Valley" was already under way.

The first company to offer me a job was a group of scientists who believed they could convince draftsmen and engineers to stop drawing by hand. Calma Corporation called its technology computer-aided design (CAD). CAD offered a way to design cheaper, safer, more efficient products by using computer workstations to create three-dimensional models instead of relying on error-prone, manual drawings.

My task was to translate confusing technical data into language the person on the street could understand. I spent many late nights with scientists from Stanford, Berkeley, and MIT trying to summarize their research into a few crisp bullet points. For me this was a dream job: Sociobiology had unwittingly equipped me to act as a bridge between science and society. After all, technology had to be married to real human needs before it would be embraced.

Computer-aided design took off as quickly as semiconductors and personal computers infiltrated Main Street. Within five years, companies around the world, from seat belt manufacturers to those building nuclear power plants, were designing products on computers. It was a radical, historic, and abrupt transformation and no one even batted an eye.

Then in a twist of fate, General Electric purchased Calma. Suddenly, Calma found itself holding the key to corporate icon Jack Welch's vision of "the factory of the future." Our new mission was to help GE successfully transition from a tired manufacturing giant to a high-tech leader in the twentieth century. I was overjoyed to report to Welch's senior staff, commonly called the *Gang of Five*, who taught me to think strategically and systemically—to look for patterns amid seemingly disparate trends.

By this time it was the eighties and Silicon Valley was booming. Computer hardware and software were changing by the nanosecond, and the Internet and cellular communications were looming on the horizon. Venture capitalists were fanning the flames of progress, and new start-ups were popping up everywhere. Before long, I succumbed to the constant, irresistible calls of recruiters on the hunt for experienced executives. I was lured from GE to a small company that introduced the world's first optical storage device; later I joined a venture that pioneered computer networking. By the time I launched my own company and began working with Apple Computer, Hewlett-Packard, Oracle, 3M, Amdahl, and Seibel Systems, novelty and complexity had become an accepted way of life. New technology was being adopted faster and faster, and no one stopped to question whether this explosive rate of change was sustainable. We all just assumed that it was.

The final piece of the puzzle arrived years later in 2006 when neuroscientists Drs. Michael Merzenich, John Kounios, and Mark Jung-Beeman independently published landmark research on how the human brain tackles complex problems. Their investigation into the conditions that lead to higher cognitive functioning were pivotal in understanding how we adapt to accelerating change and complexity.

It took all these discoveries—evolution, genetics, sociobiology, memetics, neuroscience, and my work in the vortex of Silicon Valley—to bring me to the biological reasons for the ascension and decline of civilizations.

Today, the issues that threaten human existence are clear: a global recession, powerful pandemic viruses, terrorism, rising crime, climate change, rapid depletion of the earth's resources, nuclear proliferation, and failing education. Though at first glance this list appears daunting, it is also true that we have never been in a better position to circumvent a repetitive pattern of decline.

Which brings me to the writing of *The Watchman's Rattle*.

In earlier times, ordinary citizens volunteered as watchmen to protect the welfare of their communities. They patrolled neighborhoods, lighthouses, and important institutions, watching for early signs of danger. Surprisingly, these early watchmen never carried weapons. They carried wooden rattles that made a loud, harsh, clacking noise designed to summon help. The sound of the watchman's rattle was an alarm—a call for citizens to wake from their sleep and quickly join forces against danger.

This book is the sound of the watchman's rattle in the dead of night. A summons for help. A plea to change the course of humankind by calling on the greatest weapon of mass instruction ever known: *the human brain.*

Rebecca D. Costa
Big Sur, California
April 11, 2010

I have never heard a sound beating the air,

so fraught with the spirit of trouble and need of assistance,

as the sharp crack of the watchman's rattle

reverberating in the street at the dead hour of night.

—EDWARD H. SAVAGE, 1865

— 1 —

A Pattern of
Complexity and Collapse
Why Civilizations Spiral

On the morning of August 29, 2004, I had an important insight. I remember the date because I was driving to the birth of my nephew Ben.

As I was rushing to the hospital, I was typing the address on my GPS screen, plugging my Blackberry into the cigarette lighter, getting my iPod into the docking station, plugging my laptop into a second outlet, putting my telephone headset and seatbelt on, and trying to drink my coffee. All the while keeping a two-ton vehicle moving at sixty miles an hour on the road.

That's when it hit me.

Life has become really complicated.

There was a time not long ago when all I had was a simple checking and savings account. But today, in addition to these, I find myself trying to keep up with CDs, bonds, mutual funds, REITs, ETFs, Spyders, IRAs, pensions, social security, and commodities such as oil, which seems to affect the value of every other thing. I have four different credit cards, each earning mystery points, airline miles, free rental cars, extra nights at hotels, and global discounts I can no longer keep track of. What's more, my credit card companies have now become banks and travel agencies want me to order movie tickets, make dinner reservations, and write checks through them.

And then there's the big picture.

According to the U.S. Department of State, there are forty-five active terrorist groups in the Middle East who want to kill me. On a good day I'm able to name three. And as long as I'm on a roll, let me be the first to admit that I can't figure out whether a national health care program is a good thing for me or my country because I don't have the time to study it. In fact, I don't feel qualified to vote on most of the initiatives on a ballot anymore. Although I worry about the effect that global warming and mounting government debt will have on my children's future, I no longer know what to do about them.

The world is starting to look a lot like the inside of my car.

But consider this: From an evolutionary perspective, social progress moves fast. It takes only weeks to develop the next iteration of a cell phone, law, or mortgage. And major discoveries in science, from the human genome to new fuel cells and galaxies in outer space, are also occurring at a remarkable pace. But our brains—the apparatus that must process all this new information—evolve over millions of years. So while the world is changing in picoseconds, my brain is struggling to keep up.

But what happens if it can't? Can complexity race ahead of the brain's biological capabilities?

In his book *Making Things Work,* Yaneer Bar-Yam, president of the New England Complex Systems Institute and professor at Harvard University, explains the reason complexity poses a threat: "The rule of thumb is that the complexity of the organism has to match the complexity of the environment at all scales in order to increase the likelihood of survival." He then explains why the odds feel stacked against us: "What is a complex environment? A complex environment is one that demands picking the right choice in order to succeed. If there are many possibilities that are wrong, and only a few that are right, we have to be able to choose the right ones in order to succeed."

Many more wrong possibilities than right ones? Coming from a soft-spoken physicist at Harvard, that sounds ominous.

But here's the million-dollar question: Is escalating complexity just a modern-day phenomenon? Or did people living in advanced Mayan, Roman, Khmer, and other civilizations develop the same contentious relationship with progress?

And even if they did, so what? That doesn't mean it had anything to do with their collapse.

Or does it?

A Thriving 3,000-year Civilization

Between 2600 BC and 900 AD, the highly advanced Mayan civilization spanned what are now Mexico, Guatemala, Honduras, El Salvador, and Belize. Archaeologists speculate that the empire grew to over fifteen million citizens, with population densities equal to Chicago today.

Imagine a massive, thriving society—located in the middle of one of the most hostile environments known to humankind—without any of the technology or modern conveniences we depend on today: no electricity or trucks, no telephones or police. How was order maintained for millions of people over such a vast geography? Food distributed? Garbage, sewage, and education managed?

Many people know that the Mayans were master potters, weavers, architects, and farmers, but even by modern standards the prodigious reach of the Mayan civilization represented an unfathomable leap in human achievement. Despite colossal environmental challenges, the Mayans invented a sophisticated cylindrical calendaring system, celestial charts to track weather patterns, the most advanced written language developed at that time for expressing complex ideas, and mathematics that included the revolutionary concept of "zero." They also engineered elaborate hydraulic projects that included a complex maze of public reservoirs, canals, dams, and levees. On virtually all fronts, the Mayans progressed rapidly, achieving quantum breakthroughs in technological, organizational, and artistic innovation.

Then, about a thousand years ago, sometime between 750 AD and 850 AD, the majority of the Mayan people suddenly disappeared. *In a single generation, the society collapsed.*

Why?

The prevailing theory is that the Mayan civilization met a "sudden death" because of severe drought.

Professor Gerald Haug is the best-known proponent of this view. Core samples from the Caraico Basin show that three long droughts correspond to the same time frame when Mayans abandoned their cities. "These data suggest that a century-scale decline in rainfall put a general strain on resources in the region, which was then exacerbated by the abrupt drought events."

But other scientists argue that, as malnutrition and water-born diseases became rampant, drought conditions simply worsened existing tensions between the ruling royalty and working classes. The Mayan people revolted against their rulers and began an exodus out of the cities.

Still other scholars cite the spread of a single virus as the primary cause of collapse, whereas other experts insist that food shortages merely exacerbated a history of civil war between antagonistic Mayan factions.

According to Michael D. Lemonick, an expert on Mayan culture: "Uncontrolled warfare was probably one of the main causes for the Maya's eventual downfall. In the centuries after 250—the start of what is called the Classic Period of Maya civilization—the skirmishes that were common among competing city-states escalated into full-fledged, vicious wars that turned the proud cities into ghost towns."

Renowned Yale archaeologist Michael Coe agrees with Lemonick: "The Maya were obsessed with war."

But according to Jared Diamond, author of *Collapse: How Societies Choose to Fail or Succeed* and professor of physiology and geography at UCLA, the rise and fall of civilizations can be traced to a dramatic change in environmental conditions. Diamond explains the early success of the Mayan civilization in terms of favorable water, food, temperatures, minerals, and other conditions conducive for a small society. Conversely, as the population grew and one or more of these conditions changed, progress stopped, conditions deteriorated, and citizens disbanded. In the case of the Mayans, deforestation may have started a disastrous chain of events.

From my standpoint, I agree with Diamond, Haug, Lemonick, Coe, and, for that matter, *all* theories regarding the events that led to the Mayan collapse.

How's that possible?

Simple. I am not particularly interested in *what* happened. I'm curious about *why* it happened—and whether it really happened overnight.

What occurred *before* the final event(s) responsible for the collapse of the Mayans, the Romans, the Egyptians, and the Khmer, Ming, and Byzantine empires? Did these societies adopt any behaviors—any ways of thinking—that made them *vulnerable to failure?* And if they did, are we repeating that pattern today?

Dusting Off Evolution

Today, scholars and scientists who study ancient civilizations cite environmental factors, overpopulation, wars, disease, politics, and energy and food shortages as the reasons for collapse. And although these explanations are factual, they also leave out the single most important principle of life on earth: *Evolution. The process and rate at which biological change occurs between one generation and the next.*

The principles governing the speed at which the human organism can biologically adapt offer us the single greatest insight into why civilizations succeed and fail as well as the most reliable preview of our own destiny.

Biological capabilities—the genetic features of each human organism—represent the only common denominator in every civilization and, therefore, must necessarily play a role in every civilization's outcome.

Yet, despite knowing this, when it comes to explaining how and why civilizations decline, we continue to account for every factor *except* evolution. We treat the discoveries of Charles Darwin in 1859 as if they relate only to our cave-dwelling ancestors or animals crawling around the Galápagos Islands. Evolution has become marginalized—imprisoned in the backrooms of zoology departments, treated as the precursor to microbiology, and relegated to what we see in our rearview mirrors rather than what lies ahead of us through the windshield.

Why has evolution been left out of the conversation?

Because for more than 150 years, scientists have failed to show how the principles governing evolution explain the rapid progress of human societies for a brief period, followed by their paralysis and cataclysmic failure. Somewhere along the line, biologists handed the task of understanding the relationship between evolution and modern man to psychologists and sociologists, who quickly formed theories of their own. So the ramifications of evolution on day-to-day life, public policy, and persistent, irresolvable problems were never solidified. As a result, aside from a few enthusiasts in the ecology movement and their naturalists brethren, evolutionary principles have managed to become irrelevant.

Think about it. We never hear about the effect that evolution has on solving global problems discussed on Capitol Hill, or in corporate boardrooms or in the economics, engineering, or physics departments of universities. No one mentions evolution during national elections, on television talk shows, or in a court of law unless it is in the narrow context of stirring up the tired old debate between creationism and science. We act as if evolution is something that happened in the past or to other species and, therefore, plays no role in the follies of humankind today. In short, evolution has become a *has-been*—a pair of shoes that no longer fit, an aunt who moved to another neighborhood, an old dog asleep on the back porch.

Yet, to solve the highly complex, dangerous global problems we face today, we must first recognize the crucial relationship between evolutionary change and the modern human condition. To finally answer the question scholars have wrestled with for centuries—why do human beings compulsively follow the same pattern of collapse again and again and again—we must come to terms with how we are wired to behave, irrespective of nationality, race, intelligence, wealth, or political convenience. We must look to the physiological capabilities, as well as the limitations, of the *human organism* itself.

After all, modern man has vastly different abilities than our prehistoric ancestors of just five million years ago. And given another five million years, humans will develop talents that will make our way of life today seem equally primitive. Humankind is a "work in progress," so at any point in time our biological apparatus can take us only so far.

But how far?

History makes it clear that we hit some obstacle that causes progress to slow long before the specific event(s) blamed for the collapse of a civilization—some recurring obstruction that is both natural and predictable: *The uneven rate of change between the slow evolution of human biology and the rapid rate at which societies advance eventually causes progress to come to a standstill.*

In the case of the Mayans, they became unable to "think" their way out of large, highly complex problems because they advanced to a point where traditional left- and right-brain problem-solving methods—which the human organism developed over many millions of years—were no longer sufficient to address their most dangerous threats.

Put another way, the intricacy and magnitude of the issues that the Mayans faced during their final hours—climate change, civil unrest, food shortages, fast-spreading viruses, and a population explosion—exceeded their ability to obtain facts, analyze them, innovate, plan, and act to stop them. Their problems simply became too complex.

The point at which a society can no longer "think" its way out of its problems is called *the cognitive threshold.*

And once a society reaches this cognitive threshold, it begins passing unresolved issues from one generation to the next until, finally, one or more of these problems push the civilization over the edge.

This is the real reason for collapse.

A Recurring Evolutionary Obstacle

Think of *the cognitive threshold* in this way: The rate at which the human brain can evolve new faculties is millions of years slower than the rate at which humans generate change and produce new information. So, from a strictly biological standpoint, the human brain can't help but fall behind. There is simply no way an organ that requires millions of years to adapt can keep up with change that now occurs in picoseconds.

John Stanton, CBS, ABC, and CNN commentator in Washington, D.C., and author of *Evolutionary Cognitive Neuroscience* summarizes the predicament this way:

> The world that seems so familiar to you and me—a world with roads, schools, grocery stores, factories, farms and nation states—has lasted for only an eye blink of time when compared to our entire evolutionary history. The computer age is only a little older than the typical college student and the industrial revolution is a mere 200 years old. Agriculture first appeared on earth only 10,000 years ago, and it wasn't until about 5,000 years ago that as many as half the human population engaged in farming rather than hunting and gathering.

Stanton then compares this rate of change to the pace of evolution: "Natural Selection is a slow process and there haven't been enough generations for it to design circuits that are well-adapted to our post-industrial life."

It's curious that we are willing to accept physical limitations in every other area *but* the human brain. We accept the fact that a human can't lift five thousand pounds, run a mile in thirty seconds, or stay under water for more than a few minutes. We also accept archaeological evidence that shows the human brain has been quickly evolving for the past twenty-five million years. We have museums filled with skeletal proof that our early ancestors didn't enjoy near the cognitive abilities we do today. What's more, most of us agree that the brain will continue to evolve in the future; it will adapt and mutate in response to rapidly changing environmental conditions, though no one can predict precisely how.

So, doesn't it also logically follow that we have cognitive limits today?

It seems irrational to *assume* that the left- and right-brain problem-solving methods we have evolved to this point have equipped us to address highly complex problems such as climate change, terrorism, pandemic viruses, and nuclear proliferation, especially since all these problems share one obvious characteristic: They are multilayered, chaotic issues involving many, many variables acting in dynamic ways. In fact, our problems have become so large and so complex that experts rarely agree on what the problem is anymore. As a result leaders have become completely dependent on sophisticated computer-based models—the kind used to make predictions in quantum physics—to run thousands of possible catastrophic scenarios: *What if* a dirty bomb makes it through our borders? *What if* a pandemic virus annihilates a major metropolitan area? *What if* water or food is contaminated by biological weaponry? *What if* both polar caps melt? No more simple cause and effect. No more quick diagnosis and remedy. And no more simple left- and right-brain problem-solving.

The bottom line is this: When it comes to the evolution of the human organism, it doesn't matter if we are talking about the capabilities of the brain, how fast we can run a mile, or whether we have a sufficient number of appendages to drive, talk on our cell phones, and drink a cup of coffee at the same time. Our biological capabilities determine how fast and how far we can go.

Consequently, the difference between an advanced culture that survives and one that does not may simply boil down to whether a society develops new ways to triumph over a naturally reoccurring cognitive threshold. How well do we understand our physiological limitations, our biological predispositions, and the remnants of prehistoric drives and instincts? Do we take prophylactic measures to deal with them? Or do we set aside the principles of evolution and continue to repeat an unconscious pattern of complexity and collapse?

The Early Signs

The study of early civilizations suggests that two telltale signs occur prior to the specific incident(s) blamed for their collapse.

The first sign is gridlock.

Gridlock occurs when civilizations become unable to comprehend or resolve large, complex problems, despite acknowledging beforehand that these issues may lead to their demise.

For example, we now know that the Mayans lived with drought conditions, civil war, and growing food shortages for thousands of years prior to collapse. However, foreseeing all these problems in advance was of little use. The Mayans lacked the ability to discern the complexity of their circumstances and, therefore, had little possibility of rectifying deteriorating conditions. Instead, they did what every great civilization does when it reaches a cognitive threshold: They simply passed their dangerous problems from one generation to the next as these problems continued to grow in magnitude and peril.

In this way, a cognitive threshold behaves much like a powerful undertow.

Undertows are invisible, so the only way to know we are trapped in a fatal current is when we *stop moving forward*. No matter how much effort we expend, we are unable to progress.

Similarly, the first sign that a society is at risk is gridlock. A civilization insists on deploying methods once used to resolve smaller, simpler problems to solve larger, more complex issues. Although these methods repeatedly fail, like a swimmer caught in an undertow, we stubbornly pursue variations of the same failed solutions decade after decade.

Today, we face the same cognitive threshold that the Mayans, Romans, Khmer, and other advanced civilizations once encountered. The most dangerous problems we must now solve have been with us for multiple generations. But rather than calling upon our collective resources, intelligence, and technologies to stop them, we are falling into the trap of ameliorating a few bothersome symptoms instead of implementing permanent cures. And in so doing, we dutifully pass our problems on to subsequent generations. Put another way, when it comes to climate change, terrorism, government debt, and other widespread threats, we are simply dog-paddling against the powerful undertow of our cognitive limitations.

Once again, Dr. Bar-Yam sheds light on the stubborn and dangerous nature of complex problems: "Complex problems are the problems that persist—the problems that bounce back and continue to haunt us." In other words, complex problems that cannot be solved eventually manifest themselves in the form of paralysis. Reaching an impasse is the earliest sign that a society has hit a cognitive threshold.

Then, as conditions grow more desperate, the second symptom appears: *the substitution of beliefs for knowledge and fact.*

When we are trapped in an undertow, we *believe* that if we simply step up our efforts and swim harder toward the shore, we will prevail against the current. Despite empirical evidence that this isn't working, we refuse to abandon our *belief* and persist in swimming in a direct path toward land as we grow increasingly exhausted and panic ensues. No data, information, or facts will deter us from our conviction—not even the threat of death.

Human beings are organisms that have always required both beliefs and knowledge. We drew mystical creatures on cave walls to help us capture large prey, made sacrifices to invisible forces to assure bountiful harvests, and carved idols to increase fertility. We engaged in rituals to make the rains return, gathered great armies in prayer, and practiced bloodletting for centuries to cure the ill. In fact we cannot find a single example of when humans did not embrace unproven beliefs. It doesn't matter whether we examine human societies in the deepest jungles of South America, on the remote islands off Bali, or in the most industrialized nations in the world; beliefs are a part of everyday life. So it follows that, if we cannot find a single example of a belief-free society, we must necessarily conclude that beliefs, along with the pursuit of knowledge, are just as much a part of human biology as the requirement for water, oxygen, and sustenance. *Beliefs aren't nurture; they are nature. They are not optional; they are a basic human need.*

But it is also true that, throughout history, when knowledge can be attained, we substitute facts with beliefs.

What do I mean by beliefs? Beliefs are merely ideas that have not been proven. According to Dr. James Watson, Nobel Laureate credited along with Francis Crick for the discovery of the structure of DNA, we need beliefs to function, even to cross the street: The light turns green and we need to *believe* drivers will obey the signal and stop for us. If we had no belief, we would be forced to wait until all the cars came to a complete stop before crossing the street. Without beliefs, we would have to question every assumption and action, and this would lead to enormous dysfunction. We wouldn't turn the kitchen faucet on if we didn't believe that water would come out; we wouldn't schedule a dentist appointment if we didn't believe we would be alive next week; and we wouldn't deposit money in the bank if we didn't believe it would be there when we needed it.

In this way, human beliefs are not limited to religion. We possess a wide spectrum of beliefs that help us function every minute of every day.

But we are also an organism that requires *knowledge:* proven data to make rational decisions and solve problems. There is no debate that knowledge is much harder to obtain than belief. The acquisition of knowledge requires complex cognitive processes such as abstraction, searching, learning, inference, analysis, synthesis, decision-making, and judgment. Knowledge also requires replication, application, interpretation, and scrutiny. Compared to adopting beliefs, the acquisition of facts is pricey.

A society advances quickly when both human needs—belief and knowledge—are met. In other words, we thrive when facts and beliefs coexist side by side, and *neither* dominate our existence.

But as social processes, institutions, technologies, and discoveries mount in complexity, obtaining knowledge becomes more difficult.

Suddenly, water we once fetched directly from our well comes from a faucet, and we no longer can discern where it originated, how it was processed, distributed, priced, or allocated. The same goes for our monetary system, laws, taxes, satellite television, and terrorism. Every aspect of life accelerates in complexity. Not only does the number of things we must comprehend grow, the intricacy of these things also exponentially increases. So, the amount of knowledge our brains must acquire to achieve real understanding quickly becomes overwhelming.

When complexity makes knowledge impossible to obtain, we have no alternative but to defer to beliefs; we accept assumptions and unproven ideas about our existence, our world. This is the second symptom: the substitution of beliefs for fact and the gradual abandonment of empirical evidence.

Once a society begins exhibiting the first two signs—gridlock and the substitution of beliefs for facts—the stage is set for collapse.

Trapped by the forces of an undertow, every society eventually finds itself desperately railing against cognitive limitations in a battle to reach safety until, finally, like a swimmer insisting on a direct path toward the shore, we grow exhausted and succumb.

When a civilization encounters a cognitive threshold and begins substituting beliefs for knowledge, the *specific* calamity that triggers collapse isn't far behind. Whether collapse arrives in the form of drought, a pandemic virus, or war, the real culprit is a cognitive threshold that prevents dangerous problems from being rationally understood and acted on. Facts and evidence are set aside in favor of unproven remedy, and this triggers a rapid spiral of catastrophic events.

But here's the reason the relationship between complexity and collapse is important for humankind to acknowledge at this time: *The signs of a cognitive threshold begin appearing long before collapse, so there is ample time to act.*

Looking back, scientists have uncovered a mountain of evidence that Mayan leaders were aware for many centuries of their tenuous dependence on rainfall. Water shortages were not only understood but also recorded and planned for. The Mayans enforced conservation during low rainfall years, tightly regulating the types of crops grown, the use of public water, and food rationing. During the first half of their three-thousand-year reign, the Mayans continued to build larger underground reservoirs and cisterns to store rainwater for drought months. As impressive as their ornate temples were, their hydraulic systems for collecting and warehousing water were masterpieces in design and engineering.

Sadly, conservation is a mitigating measure that often becomes confused with a lasting solution. Conservation strategies are limited because, eventually, we cannot conserve below zero: When the rains stop, there simply is no water to conserve.

As rainfall levels continued to decline, fifteen million Mayan citizens never came face-to-face with their deteriorating situation. The population was exploding, the need for water was rapidly escalating, and the annual rainfall was declining. Conservation was a good short-term strategy, but this was not the same as putting all of the society's energies toward permanently solving a problem they knew would eventually be catastrophic. Three thousand years is a long time for a civilization to implement a plethora of solutions. They could have sent out exploration parties, dug more wells, relocated large parts of the population, even constructed more reservoirs and cisterns at a faster rate. Any of these actions would have eased drought pressures. But climate change was too complex of a problem to comprehend, let alone resolve, so the Mayan civilization simply reached an impasse.

Over time, all practical solutions to complex issues such as drought, political unrest, and disease became violently suppressed in favor of worship. As the Mayans entered the second phase of collapse—the substitution of beliefs for facts—ritualistic sacrifice became the unilateral solution to *everything* threatening Mayan society. We now have indisputable evidence, unearthed in the deepest underground tunnels and caves, that mutilation, torture, and murder eventually expanded to include young women and innocent infants.

According to reporter Mark Stevenson in "Archaeologists Unearth Evidence of Human Sacrifice," "Victims had their hearts cut out or were decapitated, shot full of arrows, stoned, crushed, skinned, buried alive and tossed from the tops of temples. Children were said to be frequent victims, in part because they were considered pure and unspoiled."

For a time there were a few Mayan strongholds who resisted the temptation to succumb to beliefs alone.

One small Mayan community known as the Lamanai, located in what is now Belize, outlasted many Mayan cities by as much as three hundred years because they continued to pursue logical solutions to the drought.

Archaeologists have recently uncovered an ingenious network of underground cisterns that the Lamanai instituted to channel scarce groundwater into subterranean vessels. These vessels protected the water from evaporation and provided natural refrigeration for dwindling food supplies. There is evidence that the Lamanai continued to work their underground storage systems well into the worst periods of drought, when other Mayan cities had already turned exclusively to mystical solutions.

The Lamanai are distinguished by their continued pursuit of *both* knowledge and beliefs; they were determined to solve problems both scientifically and ritualistically rather than abandon one in favor of the other. Had the drought not persisted, it is probable that the Lamanai would have survived the great Mayan collapse.

Regrettably, as water shortages continued, the Lamanai also fell into the trap of substituting facts with beliefs. As underground cisterns grew dry, they too began using cisterns as chambers for sacrifice. Although archaeologists have determined that these drastic measures occurred much later among the Lamanai than other Mayan communities, recently unearthed human remains of mutilated women and children now provide evidence that, over time, the Lamanai followed the footsteps of other Mayan factions.

So, here's the real mystery: Rather than persist with rituals that produced no result, why didn't the Mayan Empire continue to pursue rational solutions to the drought—a problem they knew, for thousands of years, might lead to their demise?

We now know the answer.

They encountered a cognitive threshold: *As the complexity of their situation grew, the Mayans never developed complex problem-solving techniques. So, when methods designed to solve simpler issues began to fail, beliefs rushed in to take the place of knowledge.*

Then, with each inheritance, problems with drought, disease, and civil unrest grew to the size of the Mayan temples themselves, until finally, one or several of the problems conspired to eradicate a once-powerful, thriving civilization.

Which begs the question: Are we better equipped to handle a complex problem such as drought today?

Doubts about Drought

I live on the Central Coast of California, where drought is a real and growing concern.

In a recent interview with the *Los Angeles Times,* U.S. Secretary of Energy and Nobel Laureate Dr. Steven Chu expressed the urgency of the situation. "I don't think the American public has gripped in its gut what could happen. We're looking at a scenario where there's no more agriculture in California. I don't actually see how they can keep their cities going. I'm hoping the American people will wake up."

During the past decade, development has been dramatically restricted along the coast of California because existing homes scarcely have enough water to make it through the summer months. The fresh water we all depend on comes from a small amount of rainfall in January, February, and March. That's it. No matter how nicely our civic leaders paint it, that's the bottom line. Examining a fifty-year trend, it is easy to see that the situation worsens a little bit every year.

We have a time bomb on our hands, but in California we treat it the same way we treat the threat of earthquakes: We pretend it's not going to happen.

In Monterey County, where I live, there has been public debate for at least three decades over building desalination plants, drilling more wells, and diverting larger amounts of fresh water from local streams and rivers. But every year working on a permanent cure, such as desalination, is postponed in favor of additional study, and conservation efforts are stepped up beginning each spring. Economic sanctions are also seasonally put into place to discourage water use. When things get really bad, water is rationed. For six years now, we have been ordered to stop watering lawns and landscaping in July. Last year I cut my water use in half, let my yard die, and watched my water bill triple.

It was a sobering experience.

When I show up at the local water board meetings and attempt to explain why conservation is not a permanent solution but rather a short-term mitigation, everyone's eyes roll back in their heads. "Get rid of your grass and replace your landscaping with drought-tolerant plants," they say.

I really wish it were that simple. If it were, the Mayans might still be around.

Conservation buys us time, but it also delays the inevitable as root causes worsen. Regrettably, the more time we buy, the more deluded we become that we have actually solved the problem.

Successful mitigation is dangerous because it can easily be confused with a permanent cure as soon as short-term symptoms ameliorate.

Although I try repeatedly to address the larger, historical water problem on the California Coast, I am thwarted by mitigation scenarios at every turn.

I once spent three months assembling a report with charts and graphs showing our diminishing water supply by specific wells and rivers over a twelve-year period. The trends were undeniable and the solution clear: *We need to manufacture more water.* There is no possibility of conserving our way out of the problem. But "creating" drinking water is a politically controversial and difficult task. We feel much more comfortable with continued conservation even though logically we know this can't sustain us in the long run. We are swimmers flailing against the powerful undertow of a cognitive threshold.

On October 11, 2009, the *New York Times* published an article titled "California Lawmakers Again Fail to Reach Water Deal":

California lawmakers met again Saturday in hopes of reaching a deal to upgrade the state's decades-old water system but left without resolving a handful of major outstanding issues. Lawmakers offered only lukewarm reactions to their four-hour meeting with Gov. Arnold Schwarzenegger Saturday, despite the governor's claim a day earlier that they were on the verge of a historic breakthrough on water. "Sadly, there's little in the way of progress," Assembly Minority

Leader Sam Blakeslee, R-San Luis Obispo, said as he left the meeting. The state's network of reservoirs and canals dates to the term of Gov. Pat Brown in the 1960s. Schwarzenegger and many others have said the system is inadequate for today's population and the millions of people likely to be added in the years ahead.

I have been attending water board meetings in my area for half a decade. During this time, we have not generated a single drop of new water. We have the need. We have the technology. But what we do not have is the faculty to act on a complex social problem. So ideas that can solve the problem come to a slow crawl and eventually die in committee no matter how effective they may be. One day, when the water crisis is bad enough, panic will ensue.

My own modern-day experiences make me wonder if there might have been a handful of Mayans who wanted to solve a large, multifaceted problem in earnest but were also thwarted time and again—or, in their case, found themselves being sacrificed on some temple top.

Whether it's the water problem in California, global climate change, a worldwide banking crisis, gang violence, or an age-old religious war, our actions are limited to mitigating short-term symptoms. As our problems pass from one generation to the next, continuing to grow in size and difficulty, we should all be worried that one of them could have catastrophic consequences. Yet, for some reason, we continue to plod along as though these problems will fix themselves rather than worsen.

In this way, we are more like earlier civilizations than we think. We have lost our ability to understand complexity, so we are gradually surrendering to it. Similar to the Mayans, we have a natural tendency to marry simplistic beliefs and solutions to extremely complex social and environmental issues.

Dr. E. O. Wilson, biologist at Harvard University and a man *Time* magazine called one of the twenty-five most important figures of the last century, summarized this paradox in 2009 when he said, "The real problem of humanity is the following: We have paleolithic emotions; medieval institutions; and god-like technology."

Wilson sees the full picture. As the gap between the slow rate of evolution and the rapid pace of human progress mounts, the cognitive threshold is not the only obstacle we face.

Haunting Human Instincts

Evolutionary biologists widely agree that many vestiges of prehistoric human emotions, instincts, drives, and desires—on which our survival once depended—persist beyond their usefulness in the modern world. Similar to the evolution of the human brain, genetically inherited predispositions require many millions of years to adapt and mutate.

Evolution is a slow, continuous, and inaccurate process. This means that at any particular point in time, our instincts—the same biological predispositions that enabled our genetic pool to survive—are out of sync with those needed to successfully surmount modern challenges. This is the very reason adaptation and mutation occurs: *When what we have is not optimal for our environment, we change.*

Let me give just one example of how the slow evolution of instincts inhibits progress and plays a role in the collapse of advanced societies.

Over the past five million years, we have learned to respond to immediate threats extremely well. As soon as we detect danger, our adrenaline kicks in: Our bodies are flooded with stimulants and we spur into action. Biologists and psychologists call this powerful response "fight or flight" because the human physiology instantly prepares to attack or run.

This response is so strong that in extreme cases of fight or flight, humans have been known to single-handedly lift a 3,500-pound car. Journalist Josh Clark describes this astonishing reaction to an immediate threat: "In 1982, in Lawrenceville, Ga., Angela Cavallo lifted a 1964 Chevrolet Impala from her son, Tony, after it fell off the jacks that had held it up while he worked underneath the car. Mrs. Cavallo lifted the car high enough and long enough for two neighbors to replace the jacks and pull Tony from beneath the car."

The reason we respond effectively to immediate danger is due to the part of the brain called the hypothalamus. When the hypothalamus detects a threat, it sends chemical messages to the adrenal glands that activate hormones designed to create an excited, ready state. So no matter what kind of emergency we encounter, our bodies are programmed to take quick action.

In truth, the faster and more clever our ancestors' reaction to an imminent danger was, the more likely they were to survive. So it follows that we are all descendents of ancestors who ran, hid, successfully fought off, and detected threats better than others. Travis Gibbs, the author of *Renewal*, puts this another way: "Genetics simply remembered for us and passed along what had worked." Quick reactions worked, so this instinct perfected itself over time.

Consequently, today we are more effective at responding to immediate problems than we are at reacting to vague, distant problems.

We simply aren't hardwired to respond to long-range threats. When there's no immediate danger, there's no change in our body chemistry, no "fight or flight," no sense of urgency.

Nicholas D. Kristof, a columnist for the *New York Times*, describes it this way: "If you come across a garter snake, nearly all your brain will light up with activity as you process the 'threat.' Yet, if somebody tells you that carbon emissions will eventually destroy earth as we know it, only the small part of the brain that focuses on the future—a portion of the prefrontal cortex—will glimmer."

Despite all the information we have about exploiting our planet's natural resources, escalating climate change, mounting global debt, and the dangers of storing nuclear waste in the ground, we don't respond efficiently to distant problems, even when they pose a catastrophic threat. From an evolutionary standpoint, we simply haven't been around long enough to develop a response to remote threats. Sure, we can argue that we react more responsibly to social and environmental problems than the Mayans, but that comes as no surprise. With each civilization, our ability to understand and respond to long-range threats improves a little. This is the nature of evolution: Changes in our instincts and physiology are incremental, miniscule, and slow, but improvement occurs with each generation, each new civilization.

So when it comes to the uneven rate between evolution and complexity, not only does cognition fall behind, but prehistoric instincts also play a crucial role in the naturally recurring pattern of collapse. Harmful biological instincts linger for many generations, creating natural obstacles to progress.

It's disturbing to think that entire civilizations collapse and begin anew because they reach the limits of their inherited biological capabilities. And yet the history of the Mayan, Roman, and Khmer empires provide ample evidence that evolutionary obstacles eventually trigger a cascade of events that result in demise.

Complexity and the Romans

Dr. Joseph Tainter gives a compelling and realistic description of the fall of the Roman Empire in his book *The Collapse of Complex Societies*.

Like me, Tainter dismisses single-cause explanations such as the recruitment of barbarian mercenaries who weakened the Roman army, or unsound economic principles leading to scarcity of food and migration out of the cities, or the Antonine Plague (which killed as much as half the population between 165 and 180 AD). In Tainter's view, these events simply pushed a teetering civilization over the cliff, but the pattern of collapse was observable long before the final events.

Tainter argues that complexity is a greedy institution: The bigger and more intricate problems become, the more resources are needed to combat them. Eventually, a society can't amass enough resources to fend off the problems, which have gone unresolved for so long.

The example Tainter gives is the measurable decline in Roman agricultural production over several generations. As production was decreasing and populations were increasing, "the per-capita energy" began dropping at a dangerous rate. The Romans successfully mitigated this agricultural shortage for a short period by conquering their neighbors, which led to an immediate infusion of the metals, grains, slaves, and resources needed to sustain progress. As the empire grew, however, "the cost of maintaining communications, garrisons, civil governments, etc., increased. Eventually, the cost became so great that challenges such as invasions from the Huns and crop failures could no longer be solved simply by conquering more territory. At that point, the empire fragmented into smaller units."

Did the Roman Empire simply become too unwieldy? Or did the population grow too large, diverse, and demanding to be ruled by a single government?

Tainter believes that war, crop failures, disease, and political unrest may appear to have caused the fall of the Roman Empire, but in truth "diminishing returns on investments in social complexity" was the root cause. As systems for commerce, governance, and defense grew more complex, the "energy" needed to manage them simply exceeded the capabilities of the Roman people.

Then, once again, as the society encountered gridlock, beliefs began to overshadow facts and rational thinking. As large, complex problems persisted, the belief that the Roman people were a superior race grew stronger.

As the Romans became convinced of their righteous place in the world and the belief in the sacredness of Roman blood prevailed, bit by bit essential functions such as military defense were turned over to mistreated peoples. Human instincts in the form of greed, indulgence, and hedonism escalated at the same time that a dangerous dependence on foreign slaves was increasing. The Romans' belief in racial superiority prevented them from ever imagining an insurrection of inferior peoples.

Impossible!

In this way, long before the Huns invaded Rome, the conditions for collapse had already been established.

The Collapse of the Khmer

Recently, archaeologists have begun piecing together the demise of the great thirteenth-century Khmer Empire, located deep in the jungles of what is now called Cambodia. Known for the world's largest religious temple at that time, Angkor Wat, the Khmer Empire stretched over four hundred square miles and served almost a million citizens. According to author and scholar Richard Stone, Angkor represented "the most extensive urban complex of the pre-industrial world."

But today, scholars describe the collapse of Angkor society as "a cautionary tale of technological overreach." According to new archaeological findings, "Angkor was doomed by the very ingenuity that transformed a collection of minor fiefdoms into an empire."

So, what happened?

The story of Angkor eerily mimics the collapse of the Mayan Empire half a world away.

Like the Mayans, the success of the Khmer Empire depended entirely on controlling the flow and availability of water. Each year, during the wet monsoon season, water poured down from the hillsides, flooding rivers, agricultural plains, and villages. On the one hand, it was necessary to divert floodwaters to prevent catastrophic damage to homes and fertile croplands. On the other hand, these same threatening waters had to be captured so the empire could survive the drought season that also arrived each year. To manage both monsoon and drought conditions, Khmer engineers devised clever hydraulic systems that redirected excess waters to massive man-made reservoirs.

New archaeological evidence reveals that the Khmer Empire achieved unfathomable feats in hydraulic engineering: "Khmer engineers built a network of canals, moats, ponds and reservoirs . . . including a reservoir called the West Baray that's five miles long and 1.5 miles wide. To build this third and most sophisticated of Angkor's large reservoirs a thousand years ago, as many as 200,000 Khmer workers may have been needed to pile nearly 16 million cubic yards of soil in embankments 300 feet wide and three stories tall."

Today, scientists marvel at the complex network of spillways, moats, canals, reservoirs, ponds, and warning systems conceived and built by hand by Khmer engineers. There is little debate that this hydraulic technology was the basis for the success of the Empire itself.

Reliable water meant a reliable harvest, which enabled steady food supplies to be available year-round. Once food and water could be stored in surplus, the empire quickly expanded.

Then, after six centuries of successful water management, two catastrophic events occurred—one man-made and the other orchestrated by nature.

According to Dr. Roland Fletcher, archaeologist at the University of Sydney and codirector of the Greater Angkor Project, the cascade first began with a tragic engineering mistake.

Evidence indicates that engineers decided to change the course of the Siem Reap River by constructing a dam. The goal of the dam was to direct river water toward a newly built reservoir. But the engineers miscalculated and built the dam too low. So when the monsoons arrived, the dam turned into a massive spillway. Water began flowing over the top of the dam into abandoned canals, causing catastrophic damage to other parts of the system. Once the dam was fully breached and the damage had spread, the amount of water directed to other reservoirs also decreased.

Imagine the chaos, the destruction, and the dismay of the engineers who successfully built and maintained one of the most complex hydraulic systems ever created. The catastrophic failure must have been as horrifying as our modern-day experience with Hurricane Katrina. Untamed floodwaters can become a devastating, unstoppable force, whether they emanate from a modern-day levee or a poorly engineered Khmer dam.

Records show that the Khmer spent generations attempting to repair the damaged water system "that grew ever more complex and unruly." Yet year after year, the system on which the survival of the empire depended continued to deteriorate until the dam completely failed, igniting a cataclysmic chain reaction of other failures.

Then the second event struck: A series of back-to-back droughts and megamonsoons descended on the Khmer Empire between 1362 and 1392 and again between 1415 and 1440. According to Fletcher, these extreme conditions (caused by what scientists call the "Little Ice Age") "would have ruined the water system."

Similar to the fate of the Mayans, water problems bred food short-ages and disease and contributed to a malnourished Khmer army who became unable to fend off attack. There is also evidence of civil unrest and revolt.

Regardless of which final act led to the collapse of the Khmer, it is likely they would have continued to advance if the complexity of the engineering and environmental problems they faced had not paved the way for supernatural beliefs to intrude. As conditions worsened, the Khmer, like the Mayans, began putting all their faith in fetishism and sacrifices were stepped up. And once beliefs were substituted for fact, rational solutions became expendable.

Complexity and Collapse

No matter which advanced civilization we examine—Mayan, Roman, or Khmer—we find a similar pattern of collapse. In the beginning, each society overcomes insurmountable obstacles and environmental challenges. They appear to gain control over their surroundings, stabilize food and water, and build systems to assure the safety of their citizenry. Against incredible odds, innovation, diversity, and creativity all thrive. *In these societies, both beliefs and the pursuit of knowledge can be shown to peacefully coexist side by side.*

Then, over time, complexity accelerates and facts become difficult and, eventually, impossible to acquire. The society becomes unable to solve its problems, particularly those that impose no immediate threat. Then the society begins passing looming dangers from one generation to the next, as conditions worsen and survival grows more tenuous. Eventually, the society becomes dependent on short-term mitigations and unproven beliefs for remedy.

Until now, we have never understood the real consequences of the uneven rate of change between evolution and social development. It never occurred to our ancestors that they were unable to obtain the information needed to solve their most complex, dangerous problems or that beliefs would be adopted in lieu of facts and knowledge.

Today, however, we understand both the pattern of collapse and evolutionary obstacles to progress.

So we can act.

But will we?

Fortunately, there's a lot of evidence in our favor.

— 2 —

Evolution's Gift

A Breakthrough in Neuroscience

IN THE JULY 28, 2008, edition of the *New Yorker* magazine, science writer Jonah Lehrer describes how we break the cognitive threshold in his article "The Eureka Hunt: Why do good ideas come to us when they do?" Lehrer tells the story of Wag Dodge, a legendary smoke jumper from Montana. The story goes something like this:

On August 5, 1949, one of the hottest days recorded in Montana history, thunderstorms ignited a small fire just outside of Missoula in Mann Gulch.

That day, sixteen smoke jumpers led by Captain Wag Dodge left Missoula in a C-47 to extinguish a few burning acres. It was a routine mission, similar to brushfires the crew had faced hundreds of times.

When the smoke jumpers hit the ground, the fire was burning trees on one side of the gulch. But in an instant, the wind reversed itself and began blowing toward the firefighters. A violent updraft took hold. The fire quickly cut off the only access route to the river and began moving at "seven-hundred-feet-a-minute" toward the men.

Dodge ordered his crew to drop their gear and run.

The men scattered and raced toward the steep canyon walls. But when Dodge saw that the flames were less than fifty yards away, he realized the fire could not be outrun. In a moment that can only be described as "insight," Dodge made a decision to turn toward the fire, quickly lighting matches as he ran.

Remember, this was 1949. Facing a wildfire waving lit matches was the act of a suicidal maniac.

Or so it appeared.

Dodge quickly lit the grass around him, shouting frantically to the crew to follow his lead. He then crouched down in the middle of the burned perimeter. Breathing through a wet handkerchief, Dodge pulled his jacket over his neck and head and waited for the wind-whipped inferno to pass over him.

On that day, thirteen smoke jumpers perished at Mann Gulch. Only Dodge and two others who found shelter in the crevices of a canyon wall lived to tell the story of one of the worst tragedies in U.S. forest fire history.

Later, when Dodge, the oldest of the smoke jumpers, was interviewed by investigators, he was unable to explain why he spontaneously lit a backfire and laid down in the middle of it. He admitted that he had never entertained the idea before that fateful day at Mann Gulch. What puzzled Dodge and the experts was that the idea of destroying the fuel around him occurred so suddenly, even as the flames rushed to claim his life. Yet as soon as the idea appeared, Dodge was filled with overwhelming certainty that *it would work*. He never hesitated and never stopped to consider the risks. In fact, he had been so confident that he had discovered a way out, he had ordered his men to follow his example.

But what really made Dodge's insight remarkable is that no one had thought of burning a personal safety zone prior to his revelation in 1949.

As the smoke from Mann Gulch began to clear, word of Dodge's story quickly spread, and in no time burning safety zones, now known as "fire shelters," were adopted as standard training for firefighters around the world.

In 1985, forty years after the Mann Gulch tragedy, seventy-three firefighters near Salmon, Indiana, found themselves cut off from all escape routes in a wildfire that later became known as the infamous Butte Fire.

But this time the men were prepared.

They lit and took refuge in individual safety zones for more than two hours as a severe crown fire engulfed them in flames, ash, smoke, and unbearable heat. Despite being a much more intense fire than Mann Gulch, every one of the firefighters at Butte survived. Only five were hospitalized for heat exhaustion, an outcome substantially different than the sixty lives experts predicted would have been lost without Dodge's early insight.

Dodge's epiphany is as vital to firefighters who regularly risk their lives as Einstein's theory of relativity is to physicists and Darwin's breakthrough is to biologists. In each instance, the solution to a highly complex problem suddenly became simple, elegant, and discernable. And in each case their epiphanies had far-reaching ramifications.

But these were not the only common denominators.

In each case, the revelation arrived out of thin air. In each case, there was no question of whether or not the solution was correct. In each case, the IQ, cultural background, and education of the individuals widely varied. In each case, previous rules, experience, and knowledge appear to have been temporarily set aside in search of uncharted territory. And finally, in each case, the innovator was unable to explain how he arrived at his discovery; the steps leading to the breakthrough could not be retraced.

What caused Wag Dodge, a smoke jumper who lived a modest life in a small town in Montana, to have an insight that would forever change firefighting history? And what about spontaneous discoveries made by Newton, Benjamin Franklin, and Nobel Laureates James Watson, Francis Crick, Muhammad Yunus, and Charles Townes? What mechanism catapulted these individuals over seemingly insurmountable complexity, allowing them to cross an intractable cognitive threshold?

Six decades after the Mann Gulch fire, neuroscientists finally have the answer.

They've discovered that our brains work in *three* ways to solve problems. We use the left side of our brain to perform organized, deconstructive *analysis* and the right side to creatively attack problems using *synthesis*. And today we have evidence of a third, heretofore-unknown cognitive process called *insight*, a faculty uniquely designed for highly demanding, complex problems.

Left, Right, and Insight

Understanding the three methods that modern man uses to solve problems is easier when we think of them as *three stores* where the brain can shop.

In one store, the products are neatly lined on shelves alphabetically by size, color, and category. The aisles are all clearly labeled and organized in a neat, systematic way. This is the analytical left side of our brain.

When we use the left brain to tackle a problem, we collect data, organize it, and eliminate solutions one by one as we sequentially narrow our options. Then we choose a solution. The left brain uses a traceable process of analysis and deconstruction based on logic. This is how we teach children, animals, computers, and robots to think.

In the second store, instead of neat, organized rows, the products are artistically arranged in showrooms. A couch sits beside a table with a lamp across from two side chairs and a painting. There's also a coffee table with magazines, a candy dish, and a vase of flowers. This is the creative right side of our brain. It contains the same products as our left brain but is organized with flair.

The right side of the brain specializes in synthesis, working with implied information. The job of the right side is to interpret subtle clues and make creative connections between what we observe and what we know. In addition to left-brain logic, synthesis played an important role in the survival of our ancestors. As a result, our ability to process implicit information and make nonlinear conclusions has also grown very efficient. For example, the right side of the brain is extremely clever at spotting cheaters.

Why?

In prehistoric times, the ability to detect cheaters was crucial to our existence. Cheaters would steal food, lie about performing work, hoard, and hide things. These types of subversive behaviors put the rest of the tribe's ability to survive at risk. So, for many generations, the human organism has been perfecting its ability to "sense" a cheater. Even though we may not be able to articulate or justify suspicious feelings toward a stranger, our inherited instincts quickly interpret a plethora of physical clues, including gesturing, eye movement, tone in the voice, pupil dilation, and perspiration, when we encounter a cheater. These subtle signs are synthesized by the right side of the brain, immediately generating uneasy and distrustful feelings.

But let's say we catch a cheater in the act and have to determine a punishment.

Chances are we will deploy the rational, analytical left side of our brain to determine an appropriate penalty. We collect information about the law, study the facts in the case, weigh the pros and cons, and eventually narrow the options. Then we choose.

In other words, identifying a cheater may involve the synthesis of many obscure clues by the right side of the brain, but when it comes to determining justice, we often deploy the logical left brain: two different problems, two different ways our brain goes about solving them.

But what happens when the brain goes shopping and can't find an answer in either store? When a problem is too complicated for everyday left- and right-brain problem-solving?

That's where the third store comes in. It has no products on the shelves, no organized departments, no artistically arranged showrooms—just a giant empty building. In this store, we simply imagine the characteristics of the product we want, and then—presto!—it appears. This is how *insight* works.

When we use insight, we begin with an idea of what we are looking for and then allow our minds to wander—sometimes for a very brief period of time and sometimes much longer. Then, in a flash, the brain navigates through a mountain of data and scenarios and BAM!

We make connections that previously eluded us, and the precise answer we were searching for appears, only *better.*

So, returning to the example of how we detect and punish a cheater, what would an insightful solution look like?

A true and entertaining story illustrates the remarkable difference a single insight can make.

In 1974, a couple of agents working in the fraud division of the Federal Bureau of Investigation had an unusual encounter with a cheater. After spending years chasing a master forger, imposter, and con man through twenty-six countries—a colorful story depicted in Steven Spielberg's movie *Catch Me If You Can*—investigators came to the disturbing conclusion that they were on the heels of one of the most skilled tricksters ever hunted by the FBI. Just sixteen years old, Frank William Abagnale was already giving experienced agents in the United States, France, Sweden, and a dozen other nations a run for their money. What bothered agents most was that they knew the young mastermind was just getting started. With every success, it grew evident Abagnale was perfecting his craft: He was becoming smarter and more skilled, and his targets were getting bigger and more audacious. Law enforcement needed to shut Abagnale down quickly.

An international manhunt for adolescent and master forger Frank William Abagnale ensued.

Then in 1969, after Abagnale had impersonated a doctor, an airline pilot, and a university teacher; worked for the state attorney of Louisiana; and had successfully bilked banks around the world, French police finally apprehended him. After being extradited to Sweden, he was deported back to the United States, where he was quickly sentenced to twelve years in prison: an open-and-shut case.

Normally, that would have been the end of Frank Abagnale's story.

But this exceptional case led to an exceptional and ironic turn of events.

The FBI agents who had chased Abagnale for nearly a decade had a sudden insight: *Why not use a criminal to catch a criminal?*

What?

For a moment, imagine the conversation among a handful of agents—the same agents who had been outmaneuvered by Abagnale for years—pleaded for his early release in order to offer him a job inside one of the most security-conscious agencies in the world. Preposterous!

But was it really?

After serving a little more than four years of a twelve-year prison sentence, Frank William Abagnale was officially released into the custody of the U.S. Federal Bureau of Investigation. As a condition of his release, Abagnale was assigned to work with senior FBI agents to crack some of their toughest fraud cases. It turned out to be a brilliant and successful partnership.

For the next thirty-five years, Abagnale helped the FBI stay one step ahead of other fraud masterminds. Even after his caseload dwindled, he continued to teach at the FBI academy and lecture in field offices around the country. Abagnale became so fond of his new role that he started one of the most successful fraud prevention and detection companies in the world, Abagnale & Associates, Inc. Today, Abagnale's firm, headquartered in Tulsa, Oklahoma, provides security tools and consulting to more than fourteen thousand institutions, including his favorite client, the FBI.

In a surprising turn of events, a few FBI agents changed the way fraud investigators viewed their most talented targets. They proved that in certain cases, partnering produced better results than prosecuting. *Their insight turned a gifted thief from a liability into an asset.* And although some argue that there were earlier attempts by law enforcement to partner with expert safecrackers and street informers, nothing had ever been attempted on the scale of what the FBI accomplished. That's why even to this day, the story of Frank William Abagnale remains legendary among Justice Department workers.

Distinguishing an insight from a good idea isn't difficult. The solutions to complex situations arrive in the form of an epiphany. Whether it's Wag Dodge's fire shelters, the FBI's partnership with a criminal mastermind, or the apple that fell on Newton's head when he discovered gravity, a distinguishing trait of *insight* is that *game-changing ideas arrive spontaneously*—so spontaneously that neuroscientists frequently refer to insight as an "aha" or "eureka" moment.

Insight and Impasse

The term "eureka" is borrowed from the tale of Archimedes, who, according to folklore, climbed into a bathtub and, in a moment of insight, made a connection between the water spilling over the edges of the tub and the scientific principle of displacement. According to the story, Archimedes cried, "Eureka!" just as his epiphany arrived.

"Aha" and "eureka" moments frequently occur just after a person feels *stuck*. Feeling stuck simply means that the degree of difficulty of a problem is not within range of left-brain analytical problem-solving and right-brain synthesis. In other words, we've shopped and shopped in two of the three stores but have been unable to obtain the solution we need.

When we are in the middle of trying to solve a problem and feel stuck, there is no way to know whether our gridlock is temporary or permanent. That's because being stuck can mean one of *three* things:

- We don't have the information, resources, and time needed to solve a problem.
- We've reached a biological limit of what the human brain is capable of.
- A solution simply doesn't exist.

When we can't solve a problem, it's one of these reasons. But which one?

Only time can tell.

But how much time?

Sometimes we have only seconds to solve a problem, as in the case of Wag Dodge. Sometimes we have a few years, such as the agents pondering what to do with criminal mastermind Frank Abagnale. And sometimes, as with the Mayans, we have thousands of years to address a threatening environmental issue. But, no matter how much time we have, stuck is stuck.

It turns out that *when the human organism hits a cognitive threshold, it doesn't matter if we have seconds or centuries.* The magnitude and complexity of the problem to be solved simply exceed left- and right-brain problem-solving capabilities that have evolved over millions of years.

But, thankfully, the impasse can be and often is broken.

All it takes is insight, *evolution's slow correction.*

With every new civilization, the environment that the human organism must master grows more complex. New technologies develop, new discoveries are made, and more intricate systems for social order and interaction are conceived. One way to view progress is to think of each new civilization as the beneficiary of millions of man-years of knowledge generated by earlier civilizations. So, rather than start from scratch, every society stands on the shoulders of the people who came before them. This seems obvious.

What may not be obvious, however, is the fact that the laws governing evolution dictate that the human brain must, over time, adapt to an environment that is growing exponentially more complex and dangerous. The brain *must* develop new capabilities in response to the multilayered, fast-moving, over-optioned, and confusing world that now challenges it.

Without new cognitive processes, such as insight, the human brain inevitably reaches a limit in the amount of complexity it can discern. The left hemisphere of the brain becomes gridlocked because no logical systems exist for narrowing a sea of options. The right hemisphere, which specializes in the synthesis of implicit data, begins interpreting millions of obscure, unrelated facts in an attempt to make sense of a situation. We begin stringing together clues and identifying patterns that simply make no sense. Once left- and right-brain processes become gridlocked, it's a sign that we have reached a *cognitive threshold.*

So, if the cognitive threshold is responsible for the cascade of behaviors leading to collapse, all we need to break the pattern is to make certain our ability to understand and manage complexity doesn't fall behind our ability to create it. *When we develop new cognitive tools, such as insight, we can prevent a cognitive threshold from ever occurring.*

So the real question is this: Will insight evolve fast enough to solve our most dangerous problems?

Evolution and the Human Brain

Evolution moves at wildly varying speeds.

Sometimes adaptation and mutation occur very quickly and efficiently. Other times, evolution plays a protracted game of trial and error. When it comes to the human organism, we have evidence of both slow and fast, inept and efficient evolution.

To understand the pace at which biological changes occur, we first need to examine what modern genetics and paleontology reveal about the fascinating path the human organism took to become modern man.

We now know that the first important event in human evolution occurred approximately three billion years ago, when molecules in the ocean began forming simple cells. Over time, some of the cells replicated and joined forces to form more complex organisms. The transition from single-cell to multicelled organisms initiated a remarkable, prolific chain of events that led to all the life-forms we now enjoy on earth.

Then, approximately four hundred fifty million years ago, the second landmark event in human evolution occurred: We developed the features necessary to climb out of the water. Certain ocean-dwelling animals developed better mobility and thicker skin than others, and these attributes permitted them to venture onto land, where they learned to reproduce and thrive. Our earliest ancestors were among these creatures.

A third milestone happened approximately sixty-five million years ago, when many species on the surface of the earth, including the dinosaurs, suddenly perished. Scientists can't agree on *why* so many robust animals disappeared at the same time, but there is no debate about *whether* mass extinction occurred. It was during this cold, dark period that warm-blooded animals called mammals began to prosper. With so many species eradicated, competition for food and the number of predators was minimal, so conditions were ideal for mammals to thrive in terms of their size, variety, and numbers.

Then a fourth landmark event in human evolution occurred shortly after humans split from their nearest relatives, the chimpanzees: We rose up to walk on two feet. This important transformation led to a mountain of rapid physiological changes.

It was during this time, approximately four to five million years ago, that the human brain experienced its most significant change. With the development of two-legged locomotion, our brains began very quickly adapting to an avalanche of new sensory complexity.

According to Philip Brownell, renowned biologist and professor at Oregon State University, "A cascade of events occurred out of bipedalism: The freedom of hands for carrying and making things and defending from enemies. The ability to sight predators, as well as prey, from much farther away. Stereoscopy. Equilibrium. And one of the most significant changes, the explosion of the neocortex, specifically the frontal lobe responsible for processing data, abstraction, problem-solving and planning."

By Brownell's account, the attitudinal change that happened when humans went from walking on all fours to two feet caused an influx of sensory data to which the brain had to respond. What good was our newfound ability to see and smell enemies a mile away if we couldn't process the information and do something with it, such as run, hide, or attack from behind?

But sensory complexity wasn't the only change the human brain was responding to.

Brownell points to another reason for the rapid evolution of the human brain: the development of sophisticated social units. As humans formed larger communities, organization, communication, coordination, planning, cooking, fighting, maintaining order, beliefs, and customs became more elaborate. So, in addition to processing sensory complexity, the brain was also adapting to new levels of *social* complexity.

In support of Brownell, the University of Chicago Howard Hughes Medical Institute published an article in 2004 describing the research of Dr. Bruce Lahn, professor of human genetics: "Why the human lineage experienced such intensified selection for better brains but not other species is an open question. Lahn believes that answers to this important question will come not just from biological sciences but from the social sciences as well. It is perhaps the complex social structures and cultural behaviors unique in human ancestors that fueled the rapid evolution of the brain." There is substantial evidence that the environmental *and* social pressures the human brain experienced during this period were considerable, so the brain quickly began to change. But not all of the brain. Specifically, *the frontal cortex, the area that processes complexity.*

How do we know this?

Paleontologists who study human evolution point to the fast-growing "brow ridge" found in the skulls of our earliest ancestors. The protrusion of this part of the upper frontal skull was necessary to accommodate a rapidly expanding frontal cortex.

In just four to five million years, this area became one-third of the human brain. In terms of evolution, this was an extremely fast mutation—so fast that paleontologists and biologists refer to this period as a "special event in evolution" or "fast evolution."

The quick growth of the frontal cortex set the stage for the next leap in human evolution, *the development of insightful problem-solving*—a leap no less significant to the perpetuation of the human species than bipedalism.

How do we know insight is the "fifth leap"?

Although it will take millions of years of human evolution to definitively prove that the development of insight is a biological response to complexity, early research already suggests that insight is evolving.

First, insight is a cognitive process buried inside the same frontal cortex that once grew at an unprecedented speed in response to new levels of sensory and social complexity.

Second, insight is better equipped to address complexity than left-brain analysis and right-brain synthesis. Where other cognitive methods come up short, insight excels.

And third, the deployment of insight has been observed in all humans, regardless of education, culture, race, or background. This indicates that insight is a biological trait rather than a learned skill.

Science writer Jonah Lehrer recognizes the relationship between the fast evolution of the frontal cortex and insight: "Pressed tight against the bones of the forehead, the prefrontal cortex has undergone a dramatic expansion during human evolution, so that it now represents a third of the brain. While this area is often associated with the most specialized aspects of human cognition, such as abstract reasoning, it also plays a critical role in the *insight* process."

But if insight is an evolutionary response to complexity, doesn't that mean that it will take another five million years to develop? Isn't that how long it took the frontal cortex to grow?

Yes, it's true that evolution moves slowly, but it is also true that, unlike early civilizations, modern society is not entirely dependent on evolution to help us cross the cognitive threshold. We have many advantages ancient cultures did not.

For the first time in history we have a clear understanding of the pattern of collapse—a pattern that eluded previous societies. We can identify the symptoms of a cognitive threshold, and therefore we can guard against them.

We are also armed with more technology, knowledge, and options than previous civilizations.

Finally, we have discovered a third cognitive process called insight, and we are quickly zeroing in on how, why, and when this remarkable process becomes activated.

Although insight is a new discovery in neuroscience, we are now, bit by bit, deconstructing the instruction manual to one of the most powerful processes in the human brain.

For example, insight may *feel* like an accident—an unexpected epiphany—but it's really a natural, observable cognitive function.

Thanks to modern advances in magnetic resonance imaging (MRI) and electroencephalogram (EEG) technology, we can observe electrical oscillations in the brain that occur when we perform tasks and solve a wide array of problems. In recent years, this information has demystified many of the myths we have had about mental illness, learning disabilities, the relationship between aging and dementia, and even how our brains respond when we act instinctively rather than thoughtfully.

Dr. John Kounios, professor of psychology at Drexel University, and Dr. Jung-Beeman, associate professor of psychology at Northwestern University, have been using advanced MRI and EEG technology to observe how insight works in the human brain. After subjecting individuals from a variety of backgrounds to increasingly difficult problems while monitoring the electrical activity of various parts of the brain, they discovered that *insight is an identifiable and repeatable cognitive process.*

When we face a highly complex problem, there is evidence that we first attempt to use familiar dominant left- and right-brain strategies. When these strategies fail, at times our brains call on insight to tackle the problem. This leads to suddenly being able to "see connections that previously eluded [us]." In other words, insight acts like normal problem-solving on steroids: It's a lightning-fast, all-inclusive, powerful cognitive process that we are born with.

Harnessing the Power of Insight

It has become clear that simply waiting for evolution to help us cross the cognitive threshold isn't the answer. Early civilizations that met a cataclysmic end banked on primitive methods of problem-solving to rescue them from their greatest threats. However, without new cognitive abilities, none of these civilizations were able to survive the growing gap between the rate of evolutionary change and escalating complexity. Today, we find ourselves in a similar predicament. We are in gridlock. Dangerous problems such as climate change, a pandemic virus, terrorism, drug addiction, and violence grow worse with every generation, and no permanent solutions appear to be in sight.

But is it really true that we have no permanent cures?

Even though we don't know how to control insights and they happen less frequently than traditional left- and right-brain solutions, insights produce inspired results every single day.

So the first step toward breaking the cognitive threshold is to recognize an insightful solution when one happens along.

But that's not such a simple task.

For example, as I write this book, leading scientists around the world agree that a relationship exists between carbon emissions and climate change. The *extent* to which this relationship affects global warming (or cooling) is a subject of considerable debate, but in 2007 the evidence amassed by the Intergovernmental Panel on Climate Change (IPCC) connecting global warming to carbon emissions was so substantial that even the most ardent opponents softened their position.

Carbon emissions are an unfortunate byproduct of burning fuels such as coal and gasoline. Whether it's the car we drive, the factory we work in, or the air-conditioned home we enjoy, we burn coal and gas around the clock.

As a result, every industrialized country has initiated some program for moving toward environmentally friendly fuels. Some countries are more aggressive than others, but fortunately no country today, not even China, is ignoring the perils of the planet.

However, despite remarkable strides in solar, wave, and wind energy, the governments of industrialized nations are still for the most part banking on nuclear energy, specifically a new form: sodium-cooled fast neutron breeder reactors, a technology that has failed commercially for more than fifty years (the United States' Fermi I plant, Japan's Moju, and France's Phenix and Superphenix).

We need only follow the money to verify the direction in which the U.S. Department of Energy is headed. Under the Obama administration, nuclear energy was called the "big winner" in the proposed 2011 Energy Department budget. When added to the $36 billion in existing federal loan guarantees for building nuclear power plants, the total budget for nuclear energy climbed to $54 billion. What's more, this figure does not include the nuclear power's share of the $5.1 billion budget assigned to the research and development of "breakthrough technologies."

Contrast this against $3 to $5 billion in federal loan guarantees for *all* renewable energy projects, a measly $302 million budget for solar power and $123 million for wind power. Not billion, million. All told, for every dollar invested in solar, wind, and hydro energy combined, approximately $10 will be invested in nuclear power research and construction.

The fact is, when federal loan guarantees are set aside, the U.S. government budgeted almost the same amount for weatherization programs as it did for renewable energy.

The same trend can be spotted in other advanced nations as well.

As of 2010, approximately 15 percent of the power supplied to the Canadian population is generated by eighteen active nuclear reactors. In the next ten years, Canada plans to build an additional nine plants—a 50 percent increase in nuclear-generated energy.

The same goes for France, where 75 percent of the country's energy now comes from fifty-nine nuclear power plants. China has also announced plans to build *forty nuclear power plants within fifteen years* as part of their recent clean energy campaign.

In almost all countries, nuclear energy has been erroneously packaged as "clean, renewable energy" simply because it produces no carbon emissions. Although this may be true, it is also a fact that nuclear power plants produce a by-product that is far more dangerous than carbon: *live radioactive waste.*

Although not widely known, nuclear power plants must shut down approximately every eighteen months to replace their fuel rods. The old fuel rods contain short-lived, low-level poisons as well as a highly toxic, radioactive material called Np-237, which has a half-life of more than two million years. Not counting nuclear facilities already on the drawing board, today we produce the "equivalent of one-hundred double-decker buses" of nuclear waste every year—waste that has to be stored somewhere.

It turns out that nuclear energy isn't clean at all. *We're simply putting pollution into the ground instead of releasing it into the air.*

To some, burying radioactive time bombs seems like a better idea than polluting the atmosphere. To others, it feels as if we are trading one form of pollution for another and subsequent generations will eventually have to pay the price: We are heading down the same cul-de-sac we once did when we bet our short-term energy needs on oil and coal.

If ever there were a problem that needed insight, the world's growing thirst for clean energy would be it.

Recently, I watched an interview with U.S. Secretary of Energy Steven Chu where he was discussing the need for renewable energy and the Marquis Waxman bill, being debated in Congress. At one point, the interviewer confessed that he didn't understand the Marquis Waxman bill, even though he had attempted to read it several times. He asked Chu if he had any simpler ideas that could have an effect on carbon emissions and global warming—something people could actually understand and implement.

Chu smiled warmly and in a soft voice mentioned that if every roof and road were painted white, it would be the equivalent of *taking eleven billion cars off the road for eleven years.*

"It's a cheapie," he added.

Rather than build hundreds more nuclear power plants producing more radioactive waste, wouldn't we rather paint our roofs and roads white? We could practically do it overnight, and the results would be immediate. Cars would remain cooler and use less energy. Cooler roads also mean less tire wear. The sun's rays would be reflected rather than absorbed, so there would be an immediate temperature reduction around the globe. The demand for household air conditioning would drop 20 percent.

According to Art Rosenfeld, who works on the California Energy Commission and who has been advocating "cool roofs" for more than two decades, we could cut *twenty-four billion metric tons* of carbon dioxide pollution in twenty years. He explains, "That is what the whole world emitted last year." It's quick, it's harmless, and, according to *New York Times* reporter Felicity Barringer, it "represents the vanguard of a movement embracing 'cool roofs' as one of the most affordable weapons against climate change."

Now we have to ask ourselves a difficult question: Why wasn't *this* bill argued before Congress over the past six months rather than the Marquis Waxman, which only a handful of experts can understand? The current carbon reduction initiative is so complicated that it makes the tax code look easy. We will need an army of full-time lawyers to implement Marquis Waxman, and there is little that the man on the street can contribute. But everyone understands white roofs and roads.

Even more perplexing is the fact that Chu, Rosenfeld, and Barringer aren't the only ones with an insightful solution to global warming.

In their 2009 bestselling book *SuperFreakonomics*, Stephen Dubner and Steven Levitt posed an alternative view. Instead of attempting to change the behavior of all human beings on the planet, why don't we investigate what it would take to quickly *cool* the planet? When viewed from this angle, the solution looks considerably different.

The last time the earth experienced a quick decrease in temperature was during the Ice Age. Volcanic eruptions released so much debris into the atmosphere that incoming sunlight was blocked. This blockage caused the planet to rapidly begin cooling. Based on history, Dubner and Levitt propose releasing sulfur dioxide into the stratosphere to form sulfate particles that would once again block sunlight. In other words, they propose "shading the earth" to stimulate cooling. According to the two experts, this would cost approximately $250 million the first year and $100 million every year thereafter. Compare this to the estimated price tag of $1.2 trillion a year required to incrementally reduce carbon emissions—a plan that would have significantly less effect. In short, cooling the earth is simpler, cheaper, and more efficient than preventing it from getting warmer.

Like painting all the roofs and roads white, Dubner and Levitt propose an inspired, evidence-based, elegant solution that, predictably, has been met with huge resistance. Overnight, geoengineering became the poster child for man's ongoing interference with nature, and scientists and green organizations lined up to attack the notion of global cooling. Sadly, the once-pristine green movement fell into the same trap that they accused their opponents of falling into: They substituted facts with irrational beliefs and, in so doing, elevated ecology from a science to a religion.

My reason for choosing Chu and Dubner and Levitt as examples is to demonstrate that often we have powerful insights that can have an immediate impact on an extremely complex and dangerous issue such as climate change, clean water, gang warfare, or nuclear proliferation. We can have a breakthrough in technology or make a new scientific discovery that offers an effective remedy. Many of these insights feel clever, prescriptive, and brilliant in their simplicity. They are also often practical, feasible, and easy to prove.

But for some reason, we don't act.

Why don't we?

What stands in our way?

3

The Sovereignty of Supermemes

The Power of Beliefs

IN A RECENT interview, Dean Kamen, world-renowned iconoclast and inventor of the Segway scooter, explained why so many dangerous problems persist: "Whether we succeed or fail will not be based on technology. We have the technology—that's easy. Changing people's *attitudes*—that's a lot harder."

As Kamen points out, we already know what our biggest challenges are, and we also have a pretty good idea of what's needed to fix them. More than any other time in human history, we have the knowledge, resources, technology—even the spontaneous insights—needed for progress to continue.

Yet around every corner our "attitudes" seem to get in the way.

That's because from an evolutionary perspective, we aren't that different from people living in ancient times. We may have computers, airplanes, and hair dryers, but don't let that muddy the waters: The way humans think and the way we behave hasn't changed a lot. We are still trapped in essentially the same biological spacesuits, responding and processing information in primitive, often predictable ways: We still wage war against other tribes, steal each other's mates, hoard, and eat more than we need. We still pretend there is an infinite amount of clean water to drink, air to breath, and trees to cut. And we still require both beliefs and knowledge to thrive.

But how do our *attitudes* prevent us from moving forward?

The answer can be found in a little word called a *meme.*

A *meme* (pronounced like "seem") is any widely accepted information, thought, feeling, or behavior. Memes can be common sense, traditions, theories, biases, even slogans. "Don't run with scissors" is a meme, and so is rubbing sticks together to make fire. The Macarena, texting, and, in the Midwest, the practice of filling the bathtub with water when trouble is headed our way are also memes. A meme can be a passing fad that only lasts a week or an old wives' tale that lasts for centuries. Some memes are factual and some are false. Some are simple to grasp, while others are quite elaborate. And some memes are helpful, while others, such as racism, cause a great deal of harm.

Dr. Susan Blackmore, author of *The Meme Machine* and senior lecturer at the University of the West of England, describes how memes are transmitted: "Memes travel longitudinally down generations, but they travel horizontally too, like viruses in an epidemic."

Recently my five-year-old nephew reminded me of the durability of memes when I took him to the neighborhood pool. "Nana," he warned me, "you stay on the steps because you ate a sandwich. You're 'posed to wait."

Five years old and already a perfect replicator of inaccuracy.

"Wait an hour after you eat before going swimming" is a meme that has been faithfully passed from one generation to the next for over a century even though the myth of "cramping up" after eating and subsequently drowning has been scientifically debunked by experts. But that hasn't stopped this information from being passed from one person to the next as if it were a vital survival tip.

The study of memes — called memetics — provides a valuable framework for understanding how culture, knowledge, beliefs, and behaviors spread until they become an accepted way of life.

Yet, after three decades any reference to memes is still guaranteed to raise an eyebrow among the scientific community.

Controversy over Memes

The idea of memes was hijacked from Darwin. In his landmark 1976 book, *The Selfish Gene,* the father of meme theory, Dr. Richard Dawkins, professor at Oxford University, explained how Darwin's principles of Natural Selection applied to more than just the inheritance of physical traits and instincts. Dawkins argued that "units of information" called memes also compete for survival.

Like genes, through variation, mutation, and inheritance, memes increase or decrease their reproductive success as they migrate from one human organism to the next.

Dawkins was the first to define a meme as "a unit of cultural transmission or imitation."

But what exactly is a "unit"?

No one knows.

For my purposes I treat memes the same way physicists treat the existence of dark matter and quarks. In quantum mechanics it's not uncommon to uncover new laws that explain the universe by assuming a hypothetical force or object exists. With this "if only" assumption, we then go looking for the missing piece. We know the characteristics it would have, how it would behave, even where it might reside. Often, however, as in the case of dark matter, we never actually find it. So what do we do? We *pretend* it exists. And we do this because hundreds of other theories work better if things like dark matter and quarks are real.

We use this approach in other sciences too.

For example, many people are surprised to learn that Charles Darwin wrote *On the Origin of Species* long before anyone knew what a gene was. Darwin may have been missing the key ingredient needed to explain exactly how evolution worked, but this didn't prevent him from observing that *something* in nature was causing certain species to triumph and others to fail. At the time, he called such mythical *particles* that blended when animals mated "pangenesis." But Darwin's guesswork quickly folded under the scientific scrutiny of his time. It was many years *after* the publication of *On the Origin of Species* that

Gregor Mendel, a monk living in an isolated monastery in Czechoslovakia, discovered how genes really worked. Thankfully, this lack of information didn't stop Darwin from hypothesizing.

Despite Dawkins' broad definition of memes and the obvious challenge of determining measurable "units" of imitation, we can all agree on one observable fact: With every passing generation *some knowledge, behaviors, and beliefs grow stronger, some grow weaker, and some become extinct.*

For instance, with the proliferation of sexually transmitted diseases and a recent spike in teenage pregnancies, "chastity is good" appears to be gaining strength. On the other hand, with the onset of blogging, texting, YouTube, and social networking sites like Twitter and Facebook, "keep your opinions to yourself" is rapidly experiencing extinction.

But there are other, more serious memes too.

Generation after generation we perpetuate inaccurate beliefs about other religions, races, and cultures. We also continue to believe green energy is more costly than energy that damages the environment, even though this has been vigorously disproven by countless economists and corporations. We still trust an executive with an Ivy League education more than one with only street experience. We believe fast is better than slow, expensive is better than cheap, and organic is safer than other methods of farming. And many people still insist that the principles of evolution are untrue despite scientific evidence amassed for the past 150 years.

In his bestselling book, *Virus of the Mind*, Richard Brodie refers to memes as "viruses" because of their ability to rapidly infect rational thought: "Your thoughts are not always your own original ideas. You catch thoughts—you get infected with them, both directly from other people and indirectly from viruses of the mind."

It's clear that the human organism replicates memes on every subject, small and large, superficial and instrumental, correct and incorrect, for brief periods of time and for many generations, to our benefit and also to our detriment.

But what determines a meme's reproductive success?

Similar to genes, some memes become passive while others grow dominant and prevail. The difference between a meme that thrives—

becomes unilaterally accepted—and one that dies is determined by its compatibility with powerful, dominant memes called "supermemes."

From Meme to Supermeme

A supermeme is any belief, thought, or behavior that becomes so pervasive, so stubbornly embedded, that it contaminates or suppresses all other beliefs and behaviors in a society.

One way to think of supermemes is as powerful *über-editors*—beliefs and behaviors that influence everything we think and do. Supermemes grow to be unquestioned beliefs about any subject—economics, religion, justice, nature, even childrearing.

In the case of the Mayans, once they reached a cognitive threshold and became unable to solve their most complex problems, fetishism began dictating all thinking and behavior. Fetishism may have started out as a simple meme, coexisting side by side with other practical pursuits such as building reservoirs, underground cisterns, conservation, and innovative methods for food storage, but as conditions grew desperate, competing ideas fell away and were replaced by a singular, narrow viewpoint. In other words, fetishism grew from a meme to a supermeme. And once fetishism was adopted as the end-all solution to Mayan troubles, the search for alternative solutions came to an abrupt end.

But this drive toward mega-beliefs and singular solutions wasn't unique to the Mayans.

Complexity also led to pervasive beliefs in 476, following the fall of the Roman Empire, when humans entered an intolerable period known as the Dark and Middle Ages. It was during this time that order was restored and the "first sustainable urbanization of northern and western Europe" was achieved. In fact, historians consider the Middle Ages an important period in the development of modern civilization, and they point to the fact that many of the boundaries that define European countries today were established during this time.

But as the complexity of urban life in Europe grew, Christianity went from being one of many beliefs (memes) to a zealous supermeme that dictated all civic functions. Alternative ways of thinking and worship were quickly extinguished. As other ways of life were driven underground, Christian dogma grew so powerful that it culminated in the Crusades, a violent period in human history when intolerance reached new heights. Millions died as the result of religious persecution. Scientists and rational thinkers were jailed, tortured, and executed. And once again, progress came to an end.

History is full of examples in which powerful supermemes became obstacles to progress: the once staunchly held view the world was flat, the unilateral belief the sun revolved around the earth, the persistent belief that bloodletting cured more than it killed. The list goes on.

That's because once a supermeme takes hold, it becomes extremely difficult for people to imagine otherwise. This is the paradox of unquestioned, embedded beliefs: We continue believing something is true even when there is ample evidence to the contrary.

But why?

The Comfort of Beliefs

It's important to remember that *supermemes are often a response to accelerating complexity*. Über-beliefs reestablish order and meaning, so they act as a powerful tonic for what ails us.

According to eighteenth-century writer-philosopher Jean-Jacques Rousseau, "The mind decides in one way or another, despite itself, and prefers being mistaken to believing in nothing."

Rousseau was right: Humans *need* to believe. And in the face of complexity, they become less selective about what they are willing to accept as fact.

Recently, I saw how easily a meme can spread and become confused with fact: I was introduced to David J. Leinweber, financial guru and author of *Nerds on Wall Street*. He's a well-dressed, affable gentleman who can put a novel spin on just about any topic, so it's easy to see how Leinweber became an overnight sensation in an otherwise cutthroat industry.

According to Leinweber, one day, for the pure fun of it, he decided to go on a hunt for entertaining statistical relationships between the performance of the New York Stock Exchange and completely unrelated fields.

It didn't take long for him to make a bizarre discovery: Movements in the stock exchange coincided with the production of butter in Bangladesh roughly 75 percent of the time. This coincidence was so remarkable that Leinweber decided to see what would happen if he added a second trend: cheese production in the United States. Sure enough, the correlation jumped to a whopping 95 percent.

Leinweber was on a roll. So he didn't stop there.

He mixed in a third indicator, global sheep population, and, lo and behold, the three historical trends paralleled the ups and down of the U.S. stock market 99 *percent of the time.*

Amazing.

Of course, Leinweber knew these indicators were totally absurd. He embarked on the exercise: to show the dangers of relying on amusing but unrelated correlations when making financial decisions.

Leinweber described his eclectic research in this way: "It was the equivalent of finding bunnies in the clouds, pouring over data until you found something. Everyone knew that if you did enough pouring, you were bound to find a bunny sooner or later, but it was no more real than the one that blows over the horizon."

But shortly after he made his research public, something peculiar happened to David Leinweber.

The more he talked about the bizarre 99 percent correlation, the more interested mavens on Wall Street became. Large investors and portfolio managers demanded to learn more about how these relationships worked, clamoring for greater details. Overnight, Leinweber was besieged with requests to define the formula with more precision, in practical terms that could then be used to *predict* market fluctuations in the future. Leinweber became the talk of the town—he represented the next financial genius with a new way to make a fortune on Wall Street.

The exercise had turned upside down.

But how could this happen?

In every advanced civilization, beliefs trump knowledge when knowledge becomes too difficult to acquire, and Leinweber's story is just another example of this. When the financial systems governing Wall Street became too complicated to be understood even by experts, they reflexively turned to a conjurer who could simplify the problem. In other words, *it felt better to hang their hats on butter production in Bangladesh than on nothing at all.*

This is the reason, when faced with chaotic financial markets, extremely experienced investors blindly followed Bernard Madoff down a rabbit hole. It's also the reason popular analysts and personalities like Warren Buffett, Jim Cramer, and Neil Cavuto can cause a stock to rise or fall with a single, well-placed comment. Every day congregations of devotees hang on their words like scripture.

But this irrational behavior isn't limited to financial markets.

We also spend billions of dollars each year on homeopathic remedies that have been proven to have no effect. We purchase more insurance policies for home, fire, flood, cars, boats, and health than at any other time in human history, even though the odds of catastrophic loss are declining every hour in our favor. And we continue to build more golf courses in California as if the growing water shortage will one day fix itself.

What causes us to adopt irrational beliefs both as individuals and as a group?

Our vulnerability to beliefs grows stronger as our ability to acquire knowledge recedes. When faced with complexity that exceeds the biological capabilities of the brain, we become susceptible to unproven ideologies and begin acquiescing to a dangerous "herd" mentality.

The Contagion of Conforming

This brings us to the second reason supermemes spread like viruses: It is much easier to conform than to make a conscious decision about every issue, regardless of whether it's deciding the color of our roofs to the best car to drive or the most efficient way to educate our children. The more complex life becomes, the more difficult it is to acquire the knowledge we need to make a correct decision. Not only are the decisions we face more complex, we also have to make many more of them and make them faster. From this standpoint, it's no wonder that group behavior and group think are so seductive. The alternative is to become paralyzed by too much information, too many choices, and too much difficulty.

When conditions become chaotic and incomprehensible, we naturally align with the majority. We let the group decide because we believe there is special wisdom in the group's decision. The results of "group think" can be historic and disturbing, as in the cases of Nazi Germany, Mi Lai, and, more recently, the Abu Ghraib prison. But group think is by no means limited to human atrocities; it also explains the overnight sensation of the Cabbage Patch doll, the spread of Disco in the seventies, and the 2008 stampede to buy rice when news of a possible shortage leaked out.

More importantly, in a global economy conformity knows no national boundaries. Supermemes spread with lightning speed from one country to another irrespective of culture, history, and other preexisting memes.

One of the best examples of the trend toward worldwide uniformity comes from black-and-white photographer John Spence Weir, who has been visually documenting the modern history of Mexico for the past fifty years. In a recent conversation he offered a sad and haunting confession:

It turns out, the story of Mexico is really the story of the slow eradication of color. At one time all the houses were purple, pink, yellow and orange. The clothes and baskets, markets and people, were colorful, too.

But today, no one wants a purple house with yellow shutters. They all want beige. Everything is beige—beige clothes, beige stores, beige garden walls. The wealthy people wanted their homes to look just like the houses in the United States. So they started painting them beige.

If I have a regret it's that I should have been photographing Mexico in *color*, not in black and white. I missed the entire story.

It was an alarming admission from a man who spent his life capturing a nation's evaporating culture in black-and-white photographs.

The phenomenon Weir observed in Mexico may have a simple biological explanation. Experts who study human behavior speculate the drive toward uniform behavior may be a natural instinct inherited from our ancient ancestors. They suggest that survival opportunities increased when we acted as a unified group rather than as individuals. Working together enabled us to capture larger prey and to efficiently defend against more powerful predators. So, similar to jackals and wolves, our ancestors relied on the strength of the pack for their well-being. If this is true, it implies that we may be biologically predisposed to conform to the wishes and behavior of the group. This may also explain why we are naturally vulnerable to supermemes.

Regardless of whether our desire to conform is motivated by comfort, is biologically inherited, or is simply a natural inclination to take the path of least resistance, one thing is certain: When it comes to survival, singularity may be less complex than diversity, but it is also dangerous.

Singularity and Extinction

In nature, diversity exists for one practical reason: *A species that develops a broad range of characteristics and behaviors—wide diversity—increases its odds of surviving a broad range of environmental challenges.*

When changes occur, no matter what they are—drought, new predators, scarcity of food, wild swings in temperature—the odds of surviving are better for a species that has diversified than for one that has not.

In evolutionary terms, diversity acts like a genetic *insurance policy*—to guard against the complete eradication of a species. This is the reason there are more than one variety of fish, bird, and ant. They all developed different strategies to respond to the environment, so when changes occur, they have the features necessary to adapt.

Let me give you one example of how vital diversity is to survival. In the 1990s, in preparation to write this book, I sold my business in Silicon Valley and moved to an isolated, federally protected sanctuary for the endangered Smith's Blue butterfly along California's Big Sur coastline.

At the time I purchased the property, I knew almost nothing about butterflies. But shortly after moving, I had a chance encounter with Dick Arnold, the preeminent entomologist and expert on the endangered Smith's Blue. This began a chain of unplanned events.

Dick explained that, for unknown reasons, the fragile Smith's Blue was dependent on a single food source: a scraggy-looking buckwheat that grows naturally along the coast of California. During the past century, development, along with the invasion of nonnative plants, destroyed a substantial population of these buckwheat plants. But unlike other butterflies that survive on different vegetation when their favorite foods become scarce, the Smith's Blue had failed to develop diversity in their diet. So as the buckwheats disappeared, so went the population of the butterfly.

My visits with Dick Arnold, local biologist Jeff Norman, and experts from the Federal Fish and Wildlife office inspired me to begin collecting seeds from buckwheat plants on the property in an effort to propagate more plants. I funded an endowment, hired a horticulturalist to oversee planting, and retained a biologist to monitor the progress of the butterflies on the site. After eight years and constant harassment from the local water board for watering plants in violation of their "no water for landscaping" mandate, I'm happy to report the number of buckwheats on the sanctuary and neighboring hillsides has exploded. The Smith's Blue return in slightly larger numbers each year.

But there is also another, more sobering side to the story of the Smith's Blue.

From a biological standpoint, I have to acknowledge that I am artificially propping up a species that, of its own accord, reduced its food choice to a single plant. And where nature is concerned, any drive toward singularity makes a species vulnerable to extinction. It may be politically convenient to put the blame for the Smith's Blue's dwindling numbers on man's encroachment alone, but the truth is the butterfly put itself at risk by choosing not to diversify.

In nature, singularity has dangerous consequences.

But the laws governing singularity and survival aren't limited to a thimble-sized butterfly. Humans are also subject to these same principles.

It's common knowledge that diversifying our financial portfolios and company product lines is a necessary strategy for hedging against future change. We also strive for greater diversity in sports playbooks, education, department stores, and restaurant menus. In each of these instances diversity is synonymous with increasing our options, flexibility, and ability to survive.

Dr. Yaneer Bar-Yam of Harvard University exposes the critical role diversity plays as complexity grows: "A system performs well in facing complex challenges when it has *high variety*. We can understand this in the case of the modern economy and technological and corporate innovation."

A variety of memes assures a civilization's continued success. The greater the diversity of ideas, technologies, and beliefs a society has to choose from, the more likely that society can effectively respond to sudden or dramatic changes in its social and physical environment. But there's a catch: The greater the diversity, the greater the complexity.

Conversely, when a civilization shows signs of uniformity, it is an indication that society has adopted supermemes in an attempt to reduce complexity by eliminating diversity.

The Super-thwarting Power of Supermemes

Supermemes suppress diversity in the same way that chains like McDonald's, Wal-Mart, and The Gap crowd out small businesses. They homogenize and simplify choice. Whereas retail chains produce conformity in how we eat and dress, supermemes suppress variety in what we know, what we believe, and how we act.

Over time, supermemes become so widespread that they begin acting as filters through which other memes must pass, and only thoughts, behaviors, and beliefs compatible with the supermeme survive. This explains why so many insightful ideas and curative solutions have difficulty coming to fruition. It has nothing to do with the idea itself. As Dean Kamen pointed out, the real obstacles are our "attitudes"—the supermemes that drive how we think and behave.

But this doesn't mean a supermeme can't be overcome.

Remember the case of criminal mastermind Frank Abagnale? The supermeme that might have kept Abagnale behind bars was the prevailing idea of "justice" we had at the time—the Judeo-Christian belief in "an eye for an eye." Even today this ideology (supermeme) permeates every aspect of the U.S. criminal justice system, thwarting other innovative ways of managing and rehabilitating offenders.

This is the reason that Abagnale's story is so remarkable. The agents who rallied for his early release successfully overcame a deeply entrenched belief: the accepted notion that a criminal must *pay* their debt inside the four walls of a prison, grouped with other criminals, denied all of the pleasures, rights, and respect reserved for law-abiding civilians. Other productive ways for criminals to pay their debt, such as working in concert with the FBI, were simply not an option, at least not until the insights of a couple of agents challenged the existing mind-set.

Are there other examples where supermemes have led to singularity? Other beliefs and behaviors that have become so widely accepted that they have unknowingly quashed variety?

Economic Singularity

There was a time when the geographic distances between countries were all that was needed to slow or prevent the globalization of memes. But today unprecedented advances in transportation and communications allow information, beliefs, and trends to travel farther and faster than ever before.

Take economics for example.

It is fascinating how similar the economies of many industrialized countries have become in just a few decades. Almost all nations have a centralized branch of government that tightly controls interest rates, the circulation of currency, import/exports, and more. They also utilize identical economic institutions to oversee commerce—stock markets, corporate and government bonds, regulated banking, venture capital, and so forth—with very minor differences in the laws governing their financial systems. Income and purchases are taxed as the primary source of revenue for the government, and citizens are required to register and apply for a variety of licenses even to conduct everyday commerce. Sure, there are technical nuances and political differences between the economies of, say, the United States and China, but the similarities in how capital is raised, controlled, and managed as well as how economic transactions occur easily outweigh these cosmetic differences. Today, business is business, no matter where you travel.

Following the Cold War most nations adopted many of the financial institutions and processes associated with capitalism. There was a practical reason for this: *Uniformity streamlines commerce.* The global economy simply runs more efficiently when the same basic economic principles are deployed irrespective of cultural and political differences.

In short, common systems of commerce paved the way for easy and rapid economic cooperation. As a result, all nations today are striving for economic singularity, even at the expense of destabilizing global markets.

The story of my brother, Mike, demonstrates why economic uniformity between countries has so much appeal.

A few years ago at his forty-fifth birthday, Mike gathered the family together and announced he was going to start his own business.

Prior to this announcement, my brother had enjoyed a career as a successful engineer in research and development in Silicon Valley working for industry giants such as Ford Aerospace, Fairchild Semiconductor, and Lam Research. In his late thirties he met a brilliant young woman, married, and was blessed with three boys. He had a good job, healthy family, and great friends. Life was good.

Then in 2008 the economy took a turn for the worse. Mike and his wife made the difficult decision to pack up the family and move to Idaho, where their cost of living would be half what it was in California. Their spacious new home included an office above the garage that my brother immediately confiscated. Between changing diapers, fixing lunch, and driving kids to school, he began combing through personal notebooks he had kept throughout his career. They were chock-full of new product ideas he hoped to develop one day. After many months he finally settled on a small, inexpensive metal screw for four-wheel-drive recreational vehicles.

Although Mike had no experience venturing out on his own and no desire to speak or learn a foreign language, within thirty days he managed to send electronic schematics of his prototype to a dozen manufacturing companies in China. In roughly sixty days he had fabricated prototypes ready for testing. He'd also hired an overseas company to develop plastic packaging for retail sales. And now, just ninety days into his dream, he's sitting in an office over his garage in Idaho interviewing foreign distributors in the auto parts industry and negotiating license agreements with manufacturers on the other side of the planet.

The Price of Uniformity

We live in amazing times, when creating an international business has become so streamlined, even a novice can become an international concern in just a few months.

Though uniformity in global commerce is a boon for entrepreneurs like my brother, a decrease in economic diversity also has a dark side: It causes nations to be more interdependent and, therefore, more vulnerable to sudden change. Similar to the quandary of the Smith's Blue butterfly, the movement away from economic diversity toward singularity has greatly increased our likelihood of unilateral collapse.

Take global stock markets for example. During the last week in July 2007, the U.S. stock market fell almost 5 percent. It was the biggest drop since 9/11. In that same week the London FTSE plummeted 5.6 percent, the Israeli TA 25 Index dropped 7 percent, and Australia's exchange fell 3 percent. This was immediately followed by the stock markets in Hong Kong, South Korea, Tokyo, and Singapore, all losing 3 to 4 percent.

Although the size of the peaks and valleys vary between international markets, the upward and downward trends are now regularly mimicked by every exchange in the world in twenty-four-hour cycles. Virtually all global stock markets follow the same daily trend.

What accounts for the similar patterns of allegedly independent stock markets, each representing allegedly independent corporations?

If these sudden jumps in stock prices were based on rational economics, we would have to conclude that companies of all sizes, in every industry, in every country in the world all *did something during the exact same twenty-four-hour period* to cause the value of their companies to decline in unison.

Not likely. Never mind on subsequent days.

So, is there some economic principle that has the power to affect every company, currency, and economy in the world?

It turns out there are very few cataclysmic events that can alter the value of every company everywhere. The rest of the time the markets are reacting either fearfully or exuberantly according to rumors, predictions, and speculation. And as all the economies around the world have grown more uniform, the response to irrational *beliefs* has also grown similar.

There is no denying that as the economies of advanced countries begin to look alike, the dangers of contagion and global financial collapse increases. In many ways we find ourselves in the same peril as the tiny Smith's Blue butterfly: We have voluntarily narrowed our options to the point of jeopardizing our ability to survive.

In my view, the recent cascade of global financial markets that led to a worldwide recession would not have been possible one generation ago when Russia, China, and East Germany operated vastly different economies. Up until the early '70s, China maintained a rigid isolationist policy, while Russia and East Germany continued to hold steadfast to centrally controlled production, distribution, and trade. During this period of economic diversification, the downturn of one financial market had a much smaller effect on other countries because to a large extent their diversity protected them from the whimsical ups and downs of a fickle world economy. Commerce between nations may have been more challenging, but a diversity of economic systems provided a necessary safeguard against a dangerous ripple effect.

Then in the spring of 2010, the danger of economic singularity once again reared its ugly head when Greece announced that it was on the verge of defaulting on massive debt, which sent markets around the world tumbling. Suddenly the ramifications of adopting a singular currency, the euro, in order to streamline commerce became clear: Economic stability was contingent on the solvency of *every* nation who adopted the euro, so the default of one country would have disastrous consequences for the others. The members of the Eurozone, along with the International Monetary Fund, had no choice but to quickly approve a $147 billion package designed to buy Greece three years of time. It was a good mitigation, but did it do anything to address the hazards of eliminating economic diversity? Probably not.

Uniformity in economics isn't the only way countries become susceptible to contagion. As global warming, pandemic virus, terrorism, poverty, and other global problems mount in complexity, the governments of every nation now face the same *gridlock*. After all, beneath the nationality of every citizen lies a human organism, subject to the same slow rate of evolutionary change as all other life on the planet. So as the gap widens between our cognitive abilities and the magnitude of the problems we must solve, citizens in every corner of the world are beginning to substitute knowledge with look-alike überbeliefs called supermemes.

Which begs the question: If complexity begets supermemes, and supermemes beget singularity, and singularity begets extinction, how do we break the cycle?

Keeping Memes, Memes

The most important thing to understand about supermemes is this: *Supermemes are man-made, man-imposed, and man-sustained, and therefore, they can be prevented and undone.*

One way to disarm supermemes is through awareness. Once we become aware of supermemes, we can't help noticing them everywhere. Government leaders, the media, comedians, and professors begin talking about them. Conversations about unilateral beliefs show up in business meetings, around the kitchen table, on television talk shows, and at civic gatherings. Once exposed, supermemes stop working.

Author Richard Brodie acknowledges the power of awareness in this way: "People who understand memetics will have an increasing advantage in life, especially in preventing themselves from being manipulated or taken advantage of. If you better understand how your mind works, you can better navigate through a world of increasingly subtle manipulation."

Consequently, the best defense against manipulation is consciousness. The more we understand how and why memes become supermemes as well as how they halt progress, the less likely we are to allow them to govern our thoughts and become an obstacle.

A second way to eliminate repressive über-beliefs is to effect a radical paradigm shift. This can take the form of a violent cultural revolution as in the case of Cuba in the 1950s or an inspirational document such as the Bill of Rights or Darwin's *On the Origin of Species*. Sometimes a revolutionary scientific discovery breaks the momentum of pervasive beliefs: the Periodic Table of Elements, the discovery of the double helix in DNA, the Internet. Human history is filled with examples where entrenched beliefs were overthrown by an extremely powerful discovery, innovation, or army.

A third way of preventing supermemes is to eliminate the reason they develop in the first place. Supermemes emerge as a result of reaching a *cognitive threshold*. So when we develop new cognitive skills—such as insight—complexity becomes manageable. Insight defends against the dangers of pervasive ideologies by eliminating our need to adopt them in the first place.

But simply encouraging insight is not enough.

We now recognize that powerful supermemes have the ability to censor or destroy every insightful solution no matter how effective it may be. Whether it's Steven Chu's idea to paint the roads and roofs white or Dubner and Levitt's pursuit of global cooling, unproven über-beliefs prevent helpful remedies from being adopted.

In this respect, modern society finds itself in a similar position to the Mayans, Romans, and Khmer. Many insightful solutions to our most dangerous problems are already available. We have technologies, ideas, theories, innovations, inventions, and insights to solve our greatest challenges. Yet supermemes stand in our way. They form an irrational barrier between what we believe and what we know we must do.

Do we know which supermemes pose an obstacle to progress today?

It turns out we have a pretty good idea.

— 4 —

Irrational Opposition

The First Supermeme

Ⅰcame face to face with the first of five modern supermemes in 2004 when I flew to Manhattan to attend two days of back-to-back business meetings.

At the time my daughter was a student at New York University, so I was looking forward to having dinner with her and catching up on the things that interest young women venturing out on their own: boys, politics, boys, fashion, boys, music and movies I've never heard of—all followed by a long discussion about boys.

But when I arrived at the hotel I discovered a voice message saying she couldn't come.

My heart sank.

Her message was short and sweet: At the last minute some students had invited her to protest against the Iraq War, and because this was her first protest, she wasn't exactly sure how long the demonstration would last. "Go ahead and eat dinner without me. I'll get there as soon as I can," she said. Then, as if anticipating my disappointment, she quickly added, "I'm sorry Mom, but I think I have to do this—you know, speak up."

I was immediately filled with pride. I had raised a healthy, willful young woman, and this was my just desserts: civic duty first, catch up with Mom second.

So, exhausted from my cross-country commute, I grabbed a soda from the minibar, laid down on the bed, and turned on the television to watch the protestors who were, coincidentally, gathering in the streets below my room.

Within minutes—drink in hand, coat and shoes still on—I fell asleep.

When I awoke, the protestors had consumed ten square blocks of downtown Manhattan. Police officers on horses began closing off intersections as ground reinforcements armed with plastic shields waved a sea of yellow cabs and confused tourists left and right. Shopkeepers hastily locked up for the day. Media vans pulled up on the sidewalks and adjusted their cameras as frantic reporters ran toward the crowd. Then, as five o'clock rolled around, thousands of office workers emptied onto the sidewalks, adding to the confusion. A small group near a park began shouting "Stop ... the ... war!" and others joined in. As the staccato grew louder, reverberating against the towering skyscrapers, traffic choppers descended to cover the action.

This was no longer just a handful of students. It was a national news story.

Then, suddenly, in one of those wonderful, inexplicable twists of fate, I saw my daughter's face on the television.

A reporter stuck a microphone to her mouth and shouted, "Why are you here? What brought you out today?"

I sat up.

I watched as the young NYU freshman took hold of the microphone to speak on behalf of the antiwar movement. With amazing composure and conviction, my daughter stated that the original reason for going to war was the existence of "weapons of mass destruction," and now, after discovering there were no WMDs, the United States no longer had any reason to remain in Iraq. "No president has the right to retrofit reasons for war," she said.

Who was this articulate, beautiful woman on the evening news? I was again filled with admiration.

But then, to the reporter's credit, he asked a more meaningful question: "How should we leave Iraq?"

Her first response was a look of surprise. "What?"

The reporter calmly repeated his question.

"I don't know. But we *definitely* need to get out." Then, curiously, she restated the reasons she was against the war all over again. It didn't seem to bother her that she had not answered the question.

I watched in fascination as the reporter moved from one person to the next—old, young, educated, wealthy, poor, executives, and students. In almost every case, the protesters presented articulate reasons why they were against the war. However, when pressed on which withdrawal plan they favored, most of them were against any one plan and fell back to restating their opposition.

Interesting.

PEOPLE WHO PUBLICLY PROTEST are not only passionate and well meaning; they are also generally well informed. I found it incongruous that these same people advocated no real solutions. They appeared passionate about what they were against, but when it came time to propose an alternative course of action, they seemed befuddled.

It occurred to me I was witnessing some new form of gridlock: If we are against the war and also simultaneously against every withdrawal plan, how do we move forward? This, by definition, is an *impasse*.

I began wondering if this phenomenon might also be true for other complex issues. Was there some supermeme—some universal belief, value, or behavior—that was preventing us from embracing real solutions?

For years I scanned op-ed pieces, news programs, and talk shows looking for a pattern. In each venue the host, anchorperson, writer, caller, and audience members seemed adamant about what they were against, but very few people, if any, advocated a solution. In fact, every person in favor of a concrete plan was followed by twenty-five vigorously attacking it. The ratio of critics to advocates was overwhelmingly skewed. The more I listened, the more I noticed how easy it had become for people to talk about what was wrong, inaccurate, incorrect, unjust, unattractive, and impractical.

It seems we are *against* a lot of things.

For instance, most of us agree something should be done about carbon emissions. But we are also vehemently against paying higher prices for gas or being forced to purchase smaller cars. And almost everyone opposes higher taxes, yet we also expect free health care, good roads, and social security as well as unlimited police and fire protection. We are against big business bailouts, but we don't want the government to allow these companies to go under or cut more jobs. We are against handing out mortgages to people with no assets, but we also want home sales to keep rising and real estate to appreciate. Corporations demand a bigger share of the global market, but they don't want to lower prices or sacrifice short-term profits to get it. Retirees want bigger returns from their investment portfolios, but they don't want to assume any risk or pay taxes on their gains.

The list of what we don't want, what we don't agree with, and what we don't like goes on and on, easily dwarfing what it is we desire.

The Problem with "Just Say No"

When it becomes much easier to describe the things we oppose rather than the things we advocate, this indicates that opposition has grown from a meme to a supermeme.

Irrational opposition occurs when the act of rejecting, criticizing, suppressing, ignoring, misrepresenting, marginalizing, and resisting rational solutions becomes the accepted norm.

But the unilateral opposition of all remedies has dire consequences to the continuation of progress.

Remember firefighter Wag Dodge? Imagine for a moment if Dodge had dismissed his revelation that lighting a smaller fire around him would save him from a more dangerous one. Imagine if Darwin had abandoned the notion of evolution simply because he could not prove the existence of genes. Imagine if Einstein's theory of relativity had been cast aside because it wasn't consistent with accepted Newtonian physics.

Throughout human history opposition has helped to create change; but it can also be a powerful and frightening obstacle to progress.

Nothing new here.

When oppositional thinking and behavior is merely a meme, tenacity and evidence may be all that is required to allow rational solutions to prevail. *But when opposition evolves into a supermeme, solu-*

tions to our greatest threats may be prevented from coming to fruition because the resources required to overcome the opposition may simply be too great.

The powerful effect that opposition has on paralyzing progress recently hit home in yet another heated public debate that, in the end, resulted in a stalemate: no resolution, no action, no progress.

For those unfamiliar with the Central Valley region of California, this area is referred to as the "breadbasket of California." Known for its sprawling dairies, cattle ranches, and orchards, the valley is neatly flanked by two highways: Highway 99 on the east and State Highway 101 on the west. Both arteries begin in Los Angeles and run the full length of the state, forming a parallel thoroughfare through the heartland of California.

Recently, the state announced plans to build a new prison in the Central Valley.

Prison overcrowding had produced such hazardous health and safety issues for inmates and employees that Federal judges, in a frightening move, ordered the governor to release 57,000 prisoners. Overnight, the problem of growing crime bumped the latest celebrity scandal off the front page and suddenly everyone became nervous.

The governor quickly determined that the Central Valley was an ideal location to construct a new prison facility. Freeway access made it easy for prisoners to be transported, supplies to be delivered, employees to commute, and families to visit. In addition, the large areas of flat land necessary to secure a prison were comparatively inexpensive, as was labor. The infrastructure for electricity, sewage, and fire protection was also already in place. There were many good reasons for building another prison in the Central Valley.

It didn't take long, however, for the state's decision to quickly galvanize neighbors who otherwise barely spoke to one another. Practically overnight, potluck meetings were organized and petitions were distributed objecting to the construction of a prison that might put schools, families, and elderly residents in jeopardy. Other objections included the negative effect on property values, the burden on local law enforcement, and the future of nearby businesses. Almost without exception, every person living and working in the surrounding area opposed the idea of building a prison near their home.

But when asked what the state should do instead, the residents offered no alternatives. Surprisingly, many became annoyed at the question, as if the state had missed the entire point. Their point was *not to build the prison here*, and that was their *only* point. It turns out that no one was against building prisons. They were in favor of stricter law enforcement and tougher, longer sentencing. They also agreed that new facilities were needed to humanely house criminals—just not in their backyard.

But is this behavior any different from the protestors against the Iraq War who reject every troop withdrawal plan? Or politicians who oppose a public health care option but advocate no solution for caring for millions of uninsured, aging citizens? How about people who want cheap nuclear energy but don't want to store radioactive waste near their communities? Or experts and leaders who criticize every plan to stop the oil spill in the Gulf?

In my view, they all amount to the same thing: *Opposition has become the new substitute for advocacy.*

Take the problem of overcrowded prisons, for example.

There's no stronger evidence that our modern society is gridlocked by opposition than the increasing number of criminals we put on ice each year. Building more prisons is a public admission that we have abandoned all other possibilities for preventing crime. We have now surrendered to the same mitigation used by the Mayans, Romans, and Khmer: incarceration and punishment. Locking up and torturing people for their crimes isn't new. It didn't work in earlier times and it doesn't work now. Yet, we persevere as if this is some surefire way to manage society's growing crime problem.

In fact, California prisons have become so jam-packed that in 2006 the prisoners themselves decided to fight back. In a landmark case, inmates sued the State of California to limit prisoner populations. With twice the allowable capacity, they argued that the prisons presented "a dangerous hazard" to inmates and guards. The inmates won the case, and the state was ordered not to accept any more prisoners.

But for two more years, conditions deteriorated.

The state's budget crisis prevented new prisons from being built, and everything came to a standstill. Police continued to arrest offenders, prosecutors continued to prosecute, and judges continued to sentence offenders to jail—but where would they go? With prison sewage, water, and electricity services already overwhelmed, diseases were rapidly spreading, prison suicides were accelerating, and, suddenly, hard-won human rights were being ignored. Dirty, unsafe, overflowing prisons resembled stockyards, and many inmates were trapped in conditions that would disturb animal rights activists.

Then, in 2009 Federal judges finally weighed in. They ordered the governor to bring inmate overcrowding down from over 200 percent to 130 percent of capacity—a level the courts deemed acceptable, but barely.

One way to accomplish the court's order was to release prisoners who had committed nonviolent crimes or who were near the end of their sentences. Another option was to scramble for new facilities that could be used temporarily as prisons. A third alternative was to build more facilities as quickly as possible in anticipation of a growing inmate population in the future.

The governor ran with all three.

It was a courageous stance, but was it enough?

The problem with any containment strategy is that the problems eventually outgrow their containers. It doesn't matter how many prisons we build or where we build them. The fact is water is running into the bathtub faster than it can drain. This means no neighborhood—whether next door to a prison or a hundred miles away—is safe. Arguing where to build the next prison is like arguing whether the tub is filling with hot or cold water: What does it matter when the water is rushing over the top?

Which begs the question: Just how bad is the problem?

According to the Department of Justice, between 1980 and 2001 the U.S. prison population grew from approximately 396,800 inmates to 1.3 million—*threefold in the past twenty years.*

But inmate statistics alone don't paint the full picture.

Prison overcrowding has put enormous pressure on the justice system for "early release." This has shifted the overcrowding problem from prisons to probation officers, whose caseloads have now become all but impossible to manage. *Put in sobering terms, between 1986 and 2006, approximately 520 new prisoners, parolees, or probationers were added every day.* Sadly, this upward trend is occurring in other nations as well.

Per capita, we have a runaway train on our hands, and every government of every industrialized nation in the world knows it.

What's more, as the courts, governor, and private citizens direct all their attention and resources to meet the demand for incarceration, new programs aimed at prevention and rehabilitation are systematically starved of funding.

Whereas in the 1960s and '70s we were eager to provide education, career and drug counseling, and therapy to prisoners, today less than 3 percent of the prison budget in the United States is allotted for inmate rehabilitation. Most of this 3 percent is being used as seed capital to develop prison businesses that, under the guise of providing vocational training, generate profits to offset rising prison costs. Services such as one-on-one counseling have been turned over to nonprofit and religious groups in order to lower operating costs.

Yet, the more we oppose every social program aimed at prevention, the more likely crime will simply continue to grow as we pass this burden to the next generation. Opposition to every possible remedy, initiative, and insight results in nothing but gridlock. And once progress stops, collapse is not far behind.

The Illusion of Free Choice

When a society becomes oppositional, it becomes extremely easy to manipulate. Individuals who understand how opposition works become masters at swaying public opinion and negotiating favorable outcomes. In this way an oppositional culture can become vulnerable to singular ways of thinking and behaving.

When we are presented with only two choices, we often choose the less objectionable option, which, in effect, becomes decision-by-default.

Politicians, for example, are masters at using this oppositional approach.

In the United States two political parties have dominated for over 150 years: the Republicans and the Democrats. Every four years each party nominates its choice for president and puts its full resources behind campaigning for its nominee.

And, predictably, every four years the same debate over "negative" campaign tactics ensues. Candidates, supporters, and staffers all make public statements agreeing that negative campaigns must stop. At the same time it is difficult for them to ignore the fact that, across the board, negative advertisements produce the best results. You can't argue with numbers: Negative ads often give a candidate the biggest bang for their buck, especially if controversy causes the ad to be discussed for free on news and talk show programs.

So as a race gets closer to the finish line, Americans have come to expect an increase in mudslinging by organizations that are conveniently kept an arm's length away from the candidates (SwiftBoat, Moveon.org, Acorn). This allows candidates to deny any knowledge of attacks on their competitor while also benefiting from the overall effectiveness of aggressive campaign tactics.

Here we see the oppositional supermeme in full force.

The reason negative advertising works so well is this: A candidate need not win our support when all he/she has to do is simply *turn us against the only other choice.* We may feel like we're exercising free choice, but what we are really doing is opposing one candidate and, by default, throwing our support behind the only other alternative.

This is one of the reasons that for over two centuries the United States has remained gridlocked in a two-party system and why we will likely remain this way for generations to come. A two-party system is ideally suited to an oppositional society and much more efficient than a three-, four-, or five-party system. All a candidate has to do in a two-party system is find *a single reason* for us to reject an opponent to earn our vote. In this way, a two-party system is easy and economical to manage.

Here's why: During a presidential election individual states are colored blue or red depending on which candidate is most likely to win the electoral votes in that state, whereas green indicates states that are still "up for grabs." With fifty states to cover, a candidate must have some method for prioritizing where to spend his money and time. Thanks to a two-party system, candidates can focus the majority of their advertising dollars and campaign stops on states where the contest is close and pollsters have indicated the candidate has a fighting chance. This makes campaigning manageable.

Just for a moment, however, imagine the complexity and confusion if three, four, or five parties with equal size, strength, and resources were competing for our vote. Then, simply differentiating from the other candidate by turning us against him (or her) would not be enough to secure a vote. Instead, candidates would be forced to advocate tangible positions to separate from the pack. Nor could they focus on states where the race was close because with so many candidates, too many states would be too close to call. Campaigns would become hugely expensive and unwieldy because candidates would be forced to campaign in orange, yellow, lavender, and chartreuse states—wherever they might be trailing numerous competitors—a campaign manager's nightmare.

Take our most recent election—an exemplary model of opposition in action.

In 2008, candidate Barack Obama implemented a brilliant and cunning platform based on the idea of "change." At the time, George W. Bush's popularity had reached an all-time low, and so the nation was, to a large extent, already primed; they were already *opposed* to the Republican policies of the existing administration. Obama offered an opportunity to move away from these policies toward some vague and compelling alternative he called "change."

Early on, during the Democratic primary, the other popular Democratic nominee, Hillary Clinton, complained that Obama wasn't being specific about what he meant by *change*. She pointed out that *change* was just a word and demanded that he give more details. But Obama was smart. He understood better than Clinton that in an oppositional society it was suicide to provide specifics. Specifics were tantamount to painting a bull's-eye on his back because any position or program he would advocate would bring scores of critics forward. So, to Clinton's frustration, Obama remained vague. The more substance she demanded, the more Obama's speeches began to take on inspirational, motivational rhetoric. They became political sermons noticeably devoid of targets.

Obama's campaign strategists knew that all he had to do was ride the sentiment of opposition against Bush while building opposition toward Clinton. So the more specifics Clinton offered, the more targets she provided for Obama, and the country, to object to. Meanwhile, Clinton punched faster and harder at the air around her, since Obama provided nothing specific for her to strike.

Slowly, opposition to four more years of "Bill and Hillary" in the White House and four more years of "the same old Republican policies" began to build momentum. All the while Obama sparingly metered out just enough detail to signal to the nation that he was a viable option.

What's more, any attempt to oppose Obama was repackaged as opposition to the nation's *first black president.* Opposition to Obama became a subtle sign of racism, and in this way opponents found themselves trapped in a situation in which counterattacks worked against them. It was a winning trifecta that tapped into existing opposition toward Hillary, Bush, and racism.

Obama's offensive and defensive campaign strategy boiled down to a simple matter of understanding and manipulating an oppositional culture better than his opponents.

But leadership requires a person to make concrete decisions and implement real programs and legislation, so we could predict the love affair would soon sour once Obama took office. Opposing the obtuse idea of *change* may have made no sense, but opposing specifics on a public health care option, more troops in Afghanistan, auto company bailouts, and stimulus packages was easy.

Simply put, the public rose up in opposition as soon as Obama provided something concrete to oppose.

Suddenly, the same people who passionately supported Obama during the election became his greatest critics as soon as he made the decision to transfer prisoners at Guantánamo to other prisons, bail out banks, and meet with the leaders of terrorist nations. Sunday morning talk show hosts lined up to pick apart the weaknesses of the president's decisions and leadership skills. Every step Obama made in the name of *change*—an idea that once captivated an entire nation—was now met with violent resistance.

Although President Barack Obama began his term in January 2009 with unprecedented popularity in the United States and abroad, by the end of his first year in office, his approval scores had sharply declined to an average of only 50 percent. Viewed in the light of an oppositional culture, this trend was not only probable, it was inevitable.

Commercial Opposition

Oppositional behavior doesn't just work for political campaigning. It's also a highly successful way to sell products today.

Many of the recent television commercials for Apple Computer (Apple has stated they are discontinuing these commercials), for example, utilize an oppositional strategy. In them, a young man representing Apple is shown talking to a chubby, pasty, unlikeable fellow who represents a PC. The PC character is not someone whom any viewer, of any demographic, would associate themselves with. On the other hand, the Apple character is slim, hip, quick-witted, and good looking. It's easy to understand why the ad is effective: Once we disassociate from the PC character, we automatically identify with the Mac.

Although comparison advertising is nothing new, oppositional advertising is.

Comparison advertising relies on showing how one feature is better than another, but oppositional advertising doesn't present any specific features or information. It simply relies on making us dislike the only viable alternative.

Today, smart advertising executives know that when there are two perceived market leaders, they need only disparage the competition for their client to gain market share. Explaining specific advantages of a product is not nearly as efficient as pointing out the other guy's faults. It doesn't matter whether the faults are real or perceived, important or superficial. If there is an opportunity for a purchasing decision to be polarized, the next stop is negative advertising.

But knowing what we are against—no matter how strongly—is not the same as knowing what we stand for, what we want, and what we believe in. Manipulation through opposition makes a mockery of "free choice" because choices have the illusion of being free when they are anything but: A choice of one, whether default or not, is no choice at all.

An oppositional strategy polarizes choice. And choosing between two extreme options doesn't work for solving highly complex problems like global warming, war, or health care because it forces the brain into choosing "which" rather than considering "what."

Imagine if we had to examine five or ten legitimate possibilities, none of which were perfect, and the best solution was one that combined elements from each? How would decisions get made? And yet, this is what insightful thinking does. Insight rejects *"which* is best" in favor of *"what* is best." When we reframe the problem for the brain, it is amazing how different the results are.

Obama may have used the polarization of choice to his advantage during the Democratic primary and national election, but once he had secured the office of the presidency, he realized that polarization was a flawed mechanism for complex decision-making. The optimal solution was never Obama *or* Hillary. It was both, working together. Taking this logic one step further, the United States would benefit even more if superficial partisan lines could be crossed and McCain, the most experienced candidate of them all, had been offered a cabinet position. Then again, that may be asking us to take as big a leap in evolution as standing on our hind legs.

Opposition and Complexity

What causes a civilization to begin rejecting ideas, information, knowledge, and solutions across the board? Why has opposition grown into a supermeme in the twenty-first century?

All roads lead back to the cognitive threshold—that inevitable gap between the slow rate at which the human brain is capable of evolving and the rapid rate at which complexity escalates.

For example, give any five-year-old child a toy that's too complicated and watch what he does. It doesn't take long for him to push it away. Then ask him why he doesn't want to play anymore. Most of the time the answer is "I don't like it" or "I don't feel like playing anymore." It's a rare child who says "It's too hard."

The same goes for adults.

When faced with complexity, our first response is to retreat to the familiar, even if the familiar means failing. But in addition to reverting to what is familiar, we also have another reaction: fear.

We are hardwired to perceive real change as threatening, so we instinctively reject it. Sure, a few of us have the courage and tenacity to attack the complex, the unknown, and the risky. After all, this is how new discoveries are made.

But many more of us do not.

Why not?

It turns out there may be a simple evolutionary explanation for our reaction: When we choose what is familiar, we reduce danger. In nature, animals that gravitate toward what is already known and understood frequently improve their survival opportunities by lowering risk.

Conversely, when we are willing to tackle the unknown, we assume much greater risk. Though progress requires a few humans to confront danger on behalf of their group, the increased risk associated with embracing novelty is more often avoided. Therefore, through the process of Natural Selection we have evolved strong cognitive safeguards to protect us against the unfamiliar and potentially harmful.

Dr. Jeffrey M. Schwartz, research psychiatrist at the University of California at Los Angeles, has been studying the human brain's response to foreign and complex ideas for many years. During his research Schwartz uncovered a simple biological reason for opposition: a two-tiered system the brain uses to manage familiar and unfamiliar tasks.

The easiest way to understand Dr. Schwartz's research is to think of the human brain like a large factory.

Inside every factory there are routine tasks that, once mastered, demand little thought. These functions are so well defined that they are largely performed by laborers whose performance can be easily measured using simple benchmarks such as quantity and consistency.

Similarly, in the human brain familiar tasks that require almost no conscious thought are managed with great efficiency by the *basal ganglia*, the "habit center of the brain." Routine tasks such as taking a shower, buttering our toast in the morning, driving the same car on the same route to work every day—any actions that have been learned, mastered, and relegated to habit—get "shoved down" to the basal ganglia in the brain.

But, every factory also requires executives, such as the chief executive officer, to perform tasks that are less routine—responsibilities comprised of negotiating, managing crises, strategic planning, and other nonconforming functions.

The CEO's job is equivalent to the complex tasks performed by the *frontal cortex*—the same area of the human brain that began growing at evolutionary light speed when humans became bipedal and formed sophisticated social groups. This is the part of the brain that processes new information and solves difficult problems, and like the CEO, it extracts a hefty price for its abilities.

Schwartz points out that the job of the basal ganglia is to "free up the processing resources of the frontal cortex." So once tasks become habitual, the brain offloads these tasks. That leaves more horsepower for unfamiliar and complex tasks to be performed in the frontal cortex—the brain's CEO.

To illustrate how the two-tiered system works in the real world, Schwartz cites driving a car as an example. Learning to drive a vehicle requires many complex cognitive processes that quickly consume all our immediate short-term memory and demand the full attention of the frontal cortex. This is the reason, after our first driving lesson, we feel completely exhausted even though we have not been exerting any physical energy other than steering and tapping the brakes.

But once we master driving, it's another story. We drive without consciously thinking about what we're doing. Often we drive familiar routes without remembering anything about the trip despite miraculously avoiding pedestrians, other cars, running red lights, and making wrong turns. How? When driving becomes so familiar as to require no conscious thought, it can be relegated to the basal ganglia, where tasks are executed on autopilot.

But what happens when the rate of change accelerates and the environment we must navigate becomes increasingly unfamiliar and complex?

For a moment imagine driving a different route to work every day where every landmark is unfamiliar. Better yet, imagine being asked to drive an altogether different kind of vehicle and having to make several unfamiliar stops along the way while still being expected to arrive on time. In order to process this much new information, we have to fully engage the frontal cortex. Without any familiarity, it is impossible to delegate even the smallest task to the basal ganglia. In this way, accelerating complexity leads to overloading the frontal cortex—the equivalent of asking a CEO to solve a corporate crisis every second of every day.

Is it any wonder we reflexively oppose everything as fast as we can? Opposition reduces our workload and risk.

In a 2006 article, "The Neuroscience of Leadership," David Rock and Schwartz observed that "change is pain." According to the two scientists who studied the human resistance to change inside the workplace, "Much of what managers do in the workplace—how they sell ideas, run meetings, manage others and communicate—is so well routinized that the basal ganglia are running the show. Trying to change any hardwired habit (or ideas) requires a lot of effort, in the form of attention. This often leads to a feeling that many people find uncomfortable. So they do what they can to avoid change."

But the research didn't stop there. The scientists also explain why avoidance and opposition is fueled by fear.

Both point to the work of Dr. Edmund Rolls at Oxford University, who was the first to show that discrepancies between what we assume and what is reality "show up in imaging technology as dramatic bursts of light." It turns out that the discrepancies between what we *expect* to have happen and what *actually happens* are instantly reported as "errors" by the *orbital frontal cortex*, an area of the brain that is directly connected to the "fear circuitry," the *amygdala*. According to Rock and Schwartz, "The brain sends out powerful messages that something is wrong, and the capacity for higher thought is decreased. Change itself amplifies stress and discomfort."

They also observe that "the orbital frontal cortex and amygdala are among the oldest parts of the mammal brain." Therefore, the ability to report errors between what we expect and what is actually occurring helped assure the survival of our earliest ancestors.

Schwartz and Rock's important research indicates there is a natural biological resistance to anything complex. What's more, because opposition is an efficient way to reduce complexity, it comes as no surprise that oppositional behavior is accelerating. As one increases, so does the other.

Once we reach a cognitive threshold, we begin unilaterally rejecting data, ideas, and solutions in a misguided attempt to make complexity manageable.

Put another way, when our brains aren't up to the job, we simply reduce the scope of the job *to fit* our abilities. It's a dangerous form of reverse-engineering in which problems are made simpler to fit the solutions we have available—solutions that have been around for years and haven't worked.

Without cognitive tools designed to manage increasing levels of complexity, attempts to stymie opposition generally fail. In an environment where there are too many variables that are changing rapidly, the brain seeks simple explanations and fewer choices to restore order. Yet, all the while we know that the key to surviving complexity is to expand diversity and choice, not thwart them.

The Oppression of Insight

When we encourage the evolution of insight, we attack the root cause of opposition. The more we develop our cognitive capacity to manage greater complexity, the more we prevail over the compulsion to over-simplify our problems.

Schwartz put it this way: "The findings suggest that at a moment of *insight*, a complex set of new connections is being created. These connections have the potential to enhance our mental resources and *overcome the brain's resistance to change.*"

Sounds simple. Just increase insight—our brain's natural ability to process complexity—and our "resistance to change" will subside.

Except for one small problem: We now know that *opposition stymies insight.*

Beginning in 1938 with Pavlov's early experiments on the role that positive and negative reinforcement had on inducing salivation in dogs and continuing through the 1970s with B. F. Skinner's publication of *Beyond Freedom and Dignity* (or as we commonly referred to it in graduate school, *Dognity*), behavioral psychologists have been collecting irrefutable evidence that criticism, negative reinforcement, and institutionalized rigidity all inhibit creativity, productivity, and growth. Take any child who has been subjected to a critical environment and observe the results: withdrawal, low self-esteem, fearfulness, and a long list of aberrant behaviors. Both humans and animals fail to thrive in critical environments where fear of failure and rejection are fostered.

Today, neuroscientists find that the conditions necessary to stimulate insight are remarkably consistent with environments needed to encourage creativity. Studies show that relaxation, allowing the mind to wander, eliminating distractions, and positive reinforcement all encourage the occurrence of insight. Though, in fairness, these same researchers point out that insight also occurs in times of extreme duress, as in the case of Wag Dodge. The spontaneous nature of insight means that it can occur at any time, but we have mounting evidence that positive, relaxed, and creative environments may increase the likelihood of this shy form of problem-solving.

For example, almost everyone is familiar with the common practice of "brainstorming." The rules of brainstorming are simple: Once a problem is presented, everything and anything goes. The fundamental tenet in brainstorming is that no idea, comment, or solution can be criticized while ideas are being tossed around. Editing other people's suggestions is a no-no, as is peer pressure, manipulation, selling, and any other tactic that inhibits the flow of innovative thinking. In the words of chemist and author Linus Pauling, "The best way to have a good idea, is to have a lot of ideas."

Brainstorming works by creating a safe, encouraging, fun environment where the brain can *play* without fear of negative consequences.

Sadly, however, more well-intended brainstorming sessions fail than succeed due to oppositional behavior, whether it is subtle or overt. The raising of an eyebrow, the shaking of a head, or the impatient tapping of a foot may be all it takes to indicate disapproval. As a result, though the practice of brainstorming was wildly popular in the '70s and '80s, in the twenty-first century it is now largely perceived as a waste of time, unlikely to produce any significant breakthroughs.

In fact, the more troubled the economy grows, the faster exercises like brainstorming disappear, even though such out-of-the-box thinking may be one of the solutions most needed during hard times. Off-site corporate retreats, executive coaches, and team-building exercises are deemed luxuries, so they are the first to be cut from the budget. Economic pressures result in fear, rigidity, and conformity, and, concurrent with this reaction, creativity and innovation are perceived as increasingly risky.

The marginalization of innovative thinking and solutions represents one of the most dangerous effects of the oppositional supermeme. The more we oppose, the more we hinder the development of insight.

But now that we are aware of the harmful effects opposition poses, can't we just guard against it in ourselves, our children, and our country? After all, we now understand how the cognitive threshold produces opposition, how opposition paralyzes progress, and how progress is essential to stave off collapse.

Earlier civilizations didn't have this information.

We also recognize that we are biologically programmed to resist the very paradigm shifts needed to assure our survival.

They didn't know that either.

Furthermore, we have evidence that a cognitive process known as insight can unravel complexity, rendering opposition unnecessary.

Previous civilizations didn't have that going for them either.

With all of this intelligence on our side and all of our advanced technology, surely modern man need not follow the pattern that led to the demise of ancient civilizations.

But irrational opposition is only one supermeme, one man-made obstacle that stands in the way of modern progress.

We still have four more to go.

— 5 —

The Personalization of Blame

The Second Supermeme

At the end of 2009 a twenty-three-year-old Nigerian agent for Al Qaeda boarded and attempted to blow up a Northwest Airlines passenger jet on Christmas Day. The would-be bomber was quickly restrained by fast-thinking passengers and turned over to authorities as soon as the plane touched ground in Detroit.

For many, the incident reignited painful memories of 9/11, but for me it was a frightening reminder that, in the name of progress, irrational beliefs and human sacrifice continue in the twenty-first century. We're just more sophisticated about it. No more Mayan priests offering infants high atop a pyramid: Today we use airplanes instead.

That evening the president of the United States ordered an immediate investigation into the security breach.

According to the *New York Times,* in the press conference following the incident, President Obama characterized the problem as "a 'systemic failure' of the nation's security apparatus." An investigation revealed that government agencies had received advanced information from foreign intelligence sources that a Nigerian national was preparing an attack. But interagency communications broke down.

So the information never reached airport security.

Obama's assessment that a "systemic failure" had occurred was a hopeful sign. The government appeared ready to tackle a highly complex, multifaceted issue: a dangerous confluence of government protocols, technology, hate crime, diplomacy, economics, communications, biological instincts, religious beliefs, and human and civil rights. Once the president labeled the problem systemic, I, like many Americans, assumed that a systemic remedy would follow.

But instead, in a subtle about-face, Obama reversed himself.

He went looking for individual culprits.

The *Times* reported, "He said he had ordered government agencies to give him a preliminary report on Thursday about what happened and added that he would 'insist on accountability at every level.'" The more the reporters pressed, the more insistent the president became that those responsible for the failure would answer for their mistakes. As he shifted the focus to individuals, the difficult job of unraveling the cultural, technological, and territorial reasons for the failure took a backseat.

Someone had to pay.

It didn't take long for the press to begin speculating who the scapegoat would be. Janet Napolitano, Obama's secretary of Homeland Security, quickly became the lead candidate because days before the president proclaimed new security measures had failed, Napolitano had held an unfortunate press conference to announce that "the system worked." Later, Napolitano was forced to reverse herself by saying, "Our system did not work in *this* instance"— but not before Representative Dan Burton and others began calling for her resignation for "undermining the confidence of Americans."

But Napolitano wasn't the only authority to be offered up. The press briefly pointed their finger at Dennis C. Blair, the director of National Security. Then, blogs holding Secretary of State Hillary Clinton responsible began popping up all over the Internet. Someone discovered that the terrorist's father had notified the U.S. Embassy in Nigeria, warning them his son was a threat. Since all embassies report to the secretary of state, the failure of the State Department to pull the terrorist's travel visa put Clinton in the spotlight yet again.

There were other candidates as well: Leon Panetta, the new director of the CIA; Robert Mueller, the head of the FBI; and Keith B. Alexander, director of the National Security Administration, to name a few. Then Dick Cheney rallied the Republican right, and Representatives Peter King of New York and Steven King of Iowa accused the president himself for the lapse in security. Obama's "low-key attitude toward terrorism" became the catchall reason for the resurgence of attacks.

Even though Obama had originally diagnosed the security breach as a *systemic* problem, this did little to quell the public outcry for a witch hunt. Soon radio and evening talk shows, besieged with frightened callers, joined outspoken politicians. The question on everyone's mind was: Who is responsible, and why hasn't the problem been fixed?

But if the problem were truly systemic, then focusing on the errors of one or two people would hardly be helpful. Did anyone believe that removing the head of one agency—any agency—would make air travel safer? Lessen the number of attacks?

Doesn't a systemic problem require a systemic solution?

The Personalization of Blame

The administration's response to the attack on Northwest Airlines was no anomaly. In fact, many people would argue that, when it comes to politics, pinning the responsibility on one or two individuals is simply a matter of course.

And in one respect they're right. The "blame game" explains why so many of our complex problems go unfixed and then migrate from one generation to the next.

Throughout history civilizations have had a clear pattern of *foisting the responsibility for complex problems onto the shoulders of individuals whenever complex problems persist*. In fact, the larger and more dangerous the problem, the more likely that individuals are held accountable. And not just heads of state. Sometimes we blame religious leaders. Sometimes our boss, ex-wife or husband, lawyer, neighbors, doctor, parents, or broker are responsible for our troubles. And sometimes we turn the blame inward toward *ourselves* in a harsh and unforgiving way.

I call this phenomenon *the personalization of blame*, a supermeme that, like other supermemes, is a naturally occurring response to gridlock and the cognitive threshold.

The way the personalization of blame works is simple: When leaders become unable to solve complex, dangerous issues, they begin shifting the responsibility for correcting these threats to individuals. As this occurs, all the attentions, resources, and efforts required to address deeply embedded social failures are set aside in favor of persecution.

Today, however, our most persistent, most threatening, problems are *all* systemic — the unintended by-products of a tangled web of processes, social institutions, laws, technologies, behaviors, values, beliefs, traditions, and evolutionary limitations. Problems such as growing crime, population growth, ecological sustainability, recession, and pandemic terrorism are the result of many known and unknown forces acting dynamically, catalytically, and, often, randomly. Like a bag of marbles dropped on a floor, the reasons for our greatest threats scatter everywhere, with no clear pattern, and no obvious cause and effect.

Talk about complexity.

Where do we start?

Unfortunately, the truth is that we haven't yet developed efficient processes for thinking about and solving massive systemic issues. So, rather than become paralyzed by complexity, we are drawn to simpler explanations, beliefs, and behaviors instead. This includes blaming individuals for what we already know are complex, embedded problems.

Accountability Run Amuck

In the words of American business icon Robert Half, "The search for someone to blame is always successful."

Well put.

With such a reliable outcome, it's hard to resist the temptation of attributing our worst problems to individual ineptitude. Today the compulsion to assign personal blame for systemic problems is still alive and well. What's more, finger-pointing is as prevalent in business as it is in politics.

The recent crisis in the U.S. automobile industry provided a telling example of how we have become a civilization that reflexively shifts the blame for systemic issues onto individuals.

In 2008, on the verge of bankruptcy, the heads of the three largest automobile companies in the United States went calling on Congress for $34 billion in loans to "restructure their businesses." The automakers reported that thousands of jobs in manufacturing, maintenance, and parts would be lost without immediate financial assistance from the U.S. government. It was a last-ditch effort to avoid closing their doors.

The first reaction most Americans had was predictable: *opposition*. After all, these were publicly traded companies and the government had no business getting into the automobile business. People who fell victim to the oppositional supermeme vehemently objected to a government bailout while suggesting very little in the way of preventing the industry from collapsing.

But the entire U.S. economy was in a free fall—sinking into a dangerous recession triggered by massive mortgage defaults, a tumbling real estate and stock market, and a potential breakdown of the banking system. Worse yet, the United States was pulling the economies of other industrialized nations down with them. Massive layoffs were on the daily news; homeowners, businesses, and Wall Street were all scrambling to stay afloat. Given this context, the government had no choice but to seriously consider the automakers' request in spite of widespread opposition.

I, like many other Americans concerned about the downward spiral, watched as the top auto industry executives gathered in Washington to testify before Congress. Each automaker had been asked to submit accounting statements, sales projections, and loan repayment plans in advance of the televised hearings, so the live Q&A wasn't really due diligence as much as it was public relations on both sides. Even so, it was fascinating to observe how the two supermemes—

opposition and the personalization of blame—shaped the questions, the testimony, and, ultimately, the outcome of a national drama.

In addition to being opposed to lending taxpayer money to bail out private corporations, the congressional panel made repeated allegations that the auto executives *themselves* were to blame for losing global market share and driving their companies towards bankruptcy. Essentially, Congress wanted to know why taxpayers should bail them out when their situation was of their own making. The once-powerful auto executives were suddenly being treated as if they had come asking for money to help save the family farm rather than rescue a hundred-year-old industry and staple of the American economy.

Then, out of nowhere one of the congressmen asked the auto executives how many of them had flown to the hearings on a private jet.

The executives looked bewildered. All of them raised their hands in unison and appeared confused as to how the question was relevant to the loans they needed.

From there it took only minutes for the photograph of the executives raising their hands to go viral. With news that they had flown in private planes to Washington, D.C., every politician, political commentator, talk show host, reporter, and citizen suddenly objected to helping rich CEOs who don't fly commercial like the rest of us.

But the collapse of the auto industry was hardly the fault of a handful of men flying around in private jets. Blaming these individuals for the failure of an entire market didn't make sense. Yet, it quickly became "us" versus "them," as people began associating the fleet of private planes with a spendthrift mentality. Rather than rallying to help a collapsing industry and preventing the loss of millions of jobs, the nation aimed their frustrations toward three baffled executives.

Then, just three months later, the CEO of General Motors, Rick Wagoner, became the predictable scapegoat. The president of the United States fired him, and U.S. citizens breathed a sigh of relief as they watched a wealthy, executive get his just desserts. Few would argue that, in the middle of a global financial crisis, three months is insufficient time to assess whether Wagoner could turn GM around, but that's not the point. It made for good public relations to blame someone for the auto industry's woes, and in this instance, Wagoner's number came up.

Although it may have felt good to pinpoint the blame on one person, the truth is that the auto industry had been hit by the same recessionary forces that destroyed the housing market, had all major retailers reporting significant losses in sales, and froze credit and stock markets around the world. These same recessionary forces dealt a final blow to an industry already weakened by foreign competition, costly union contracts, rampant gas prices, and antagonistic tariff and import/export policies.

Even if you believe the auto industry deserved their fate because they didn't develop fuel-efficient cars fast enough, it is likely the industry may have survived to fight another day had an unprecedented global recession not driven consumers away. However, rather than considering the plethora of reasons the automobile manufacturers were in trouble, it was much simpler to *personalize the blame.*

Our misguided punitive attitude toward the auto industry executives almost resulted in a refusal to bail out a vital part of the American economy that would have triggered the loss of millions of jobs. This is the problem with personalizing blame: Though it may feel good in the short run, it does nothing to fix the core, systemic issues. In fact, it can often exacerbate our challenges, piling on more negative consequences to an already indiscernible mess.

Obscuring the Facts

The nation had a similar reaction when it learned that AIG had paid bonuses to their executives after receiving a helping hand.

Employment contracts obligating AIG to pay bonuses to management were in place long before the U.S. government was called on to bail the failing giant out of pending bankruptcy. Nonetheless, public outrage quickly grew over distributing tax dollars to the very executives *accountable* for the company's failure. Like Rick Wagoner, the new AIG CEO was called to testify in front of Congress. In this case, however, the CEO had wisely accepted his post at AIG for an annual salary of just one dollar a year. This, and this alone, saved him from being single-handedly blamed for AIG's troubles.

Smart fellow.

Still, when you look at the AIG bonus debacle from the standpoint of John Q. Public, it's easy to understand why the country became incensed. Record numbers of Americans were losing their homes and jobs, retirees who had saved money all their lives suddenly no longer had enough money to retire, and people had to cut back on insurance, medications, and doctor visits. Families questioned whether they could send their children to college, buy new tires for their cars, or pay their credit card bills. In this climate, word of any kind of windfall came as an angry blow.

But independent of our beliefs and personal feelings, what *facts* did we really have about the bailout?

According to Anne Szustek of the Associated Press, AIG received approximately $153 *billion* in financial aid from the U.S. government. U.S. citizens became outraged when they discovered that approximately $165 *million* of this money would be paid as bonuses to AIG executives—the very people we considered "responsible" for the company's financial problems. But the truth of the matter is that the controversial bonuses that everyone, including the president, focused their attention on represented about *one-tenth of one penny for every dollar loaned to AIG*. In light of these facts, it is incomprehensible how such an insignificant amount of money could have so easily derailed an entire country. In earlier times we would have recognized this as a "red herring," but today we blindly rush to blame a few key individuals for the collapse of entire business sectors, and in that process, we turn ourselves into the "victims" of evil-doers.

In much the same way, it is now common practice to blame the actions of foreign nations on a single figurehead. It is convenient to believe that our international conflicts would disappear if certain leaders were simply removed. We imagine the impetus behind Al Qaeda is Osama bin Laden, and this leads us to believe that when we capture him, the organization will be disarmed and the world will be a safer place. Yet, the facts reveal that the strength of Al Qaeda is in its globally dispersed, highly autonomous sleeper cells. Cutting off just one head of a Hydra isn't likely to stop attacks like the Northwest Airlines incident; it's much more likely to cause two more heads to grow (as Greek mythology tells us).

That doesn't stop politicians from also insisting the impetus behind the North Korean rocket launches is Kim Jong Il. We believe that if we offer him as much attention as we give movie stars, and reinstitute trade, he will stop his pursuit of nuclear weapons.

We also believe our greatest threat to nuclear technology in the Middle East is Mahmoud Ahmadinejad of Iran. If neighboring countries would put sufficient economic pressure on him, he might stop building nuclear power plants. Yet, year after year these countries continue trading with Iran, hoping one day to acquire the technology for their own nuclear programs.

We are quick to label leaders who act in ways different from us as lunatics, immoral, misguided, or ill informed, and in this way, we pin the blame for massively systemic socioeconomic issues on specific personalities. It's so much easier to hold an individual accountable than to address complex problems that have plagued human civilizations for centuries. It's so much easier to believe that all we have to do is change one person's mind or simply eliminate him. It's so much easier to substitute unproven beliefs for facts.

But political heads of state, like automobile executives and the CEO of AIG, represent the will of larger cultures. Blaming *one person* for our troubles is akin to blaming the Pope for a shortage of parking spaces at the local church. Consider how many leaders have become the head of state in Israel, Iraq, Yemen, Pakistan, Yugoslavia, Russia, China, and the United States during just our lifetime? Have any of them been successful at solving the most dangerous systemic problems that face humankind? Or have global problems continued to persist and grow in magnitude regardless of who takes the throne?

Despite this, year after year we pin our hopes and our blames on one-dimensional solutions and charismatic leaders who point fingers at each other as if this were the reason for systemic gridlock. Yet the problems we now face are massively complex and are far beyond the remedy of any single human being. Like us, our leaders no longer have the biological capabilities to solve massively complex problems.

And that isn't their fault.

Blinded by Self-Recrimination

Not surprisingly, the personalization of blame extends far beyond the leaders of nations and public corporations. When a civilization reaches a cognitive threshold, where the complexity of its problems exceeds its cognitive abilities, the responsibility for fixing difficult social issues is foisted on ordinary citizens as well. Despite acknowledging that millions of people are suffering from the same affliction, it is easier to blame individuals for their failure than it is to face embedded systemic pressures. So our toughest problems such as obesity, depression, and addiction are repackaged as personal tribulations that each individual must overcome.

Today, the rationale for personal responsibility goes something like this: We as individuals make choices, and these choices have consequences. When we choose poorly, the result is misfortune. When we choose wisely, we succeed at work, love, and life. Behave responsibly, and riches and happiness are ours. Behave poorly, and the result is failure followed by despair. Personal accountability has become the new mantra of the twenty-first century. People everywhere—self-help gurus, politicians, doctors, teachers, parents, psychiatrists, law enforcement, and executives—preach personal responsibility and self-empowerment as the cure for all our troubles, large and small.

Not surprisingly, recent data released by Marketdata reveals that the self-help industry has been growing steadily by almost 10 percent per year since 2000 and is now an $8 billion business in the United States alone.

That's right: We spend $8 billion a year to tell us how to solve our own problems.

According to self-help experts, all of our problems are simply a matter of cause and effect. It doesn't matter what the specific problem is, whether it's credit card debt, overcoming drug addiction, global warming, crime, obesity, a bad marriage, or a worldwide recession. We made our bed and, therefore, we can choose to lie in it or not. By following a few easy steps, we each have the power to conquer our greatest obstacles and manifest the life we envision.

Sounds good.

But is this really true? Is the solution to overcoming mounting debt and obesity just a matter of making better choices?

Almost everyone I know is recycling and trying to pay off their loans. We're making conscious efforts to lower our cholesterol, eat fewer sweets, save money, and stop smoking. We seek out therapy and consume more mood-adjusting pharmaceuticals than at any other time in history. We also want world peace, a safer place for our children to grow up, and clean, renewable energy, and we make many personal sacrifices to have these things.

So, doesn't it seem logical that large social problems like poverty, climate change, and consumer debt would gradually subside as we become more responsible individuals, as we each fix our small share of the problem? Isn't personal empowerment the best way to incrementally solve our greatest challenges?

You would think so.

But it turns out that there are limitations to personal accountability.

The Little Engine That Can't

No matter how hard individuals swim upstream, the progress they can make is proportional to the strength of the current, and very often, that current is simply too powerful to overcome. Large systemic issues aren't just a matter of individual responsibility. Our persistent personal problems are the result of rapidly escalating complexity, a naturally occurring cognitive threshold, inherited biological predispositions, and powerful supermemes that all conspire to work against free will.

Today, the resistance every individual must overcome to solve their "personal" problems is nothing short of colossal.

In an article titled "Using Willpower to Your Advantage," columnist Allie Firestone summed up the challenge: "I'm all about making life-changing promises; in fact, I do it quite frequently. The problem is, all too often, the *old me* (sans weekly yoga classes, control over my finances, and a cleaner apartment) resurfaces before I have time to even appear to stick with my resolution."

This is what happens when individuals become confused over which problems are systemic and which are personal. We paint an incomplete and unjust picture when we blame ourselves for every difficulty because, in so doing, we deny the very existence of systemic problems.

The thinking is that if everything is up to the individual, then there is nothing to fix but individual behavior.

So solutions to complex issues such as recycling to stop the depletion of the world's natural resources become just a matter of each of us putting our glass, cardboard, and plastic refuse into colorful bins once a week. If everyone does their part and recycles, the problem will be corrected or, at the very least, substantially slowed. By accepting individual responsibility, we each become part of the solution.

If only it were that easy.

Don't get me wrong—I'm a big recycler myself.

Having said that, I also admit that on Friday mornings, when I'm diligently loading up my colored crates with empty bottles and old newspapers, I can't help but wonder why I do it. I wonder about the impact I and my well-intentioned neighbors are really having and worry about the trade-off between recycling and the second garbage truck they now send around to collect our heaping mass of soda cans. What about the gas, carbon emissions, and energy required to manufacture and maintain the second truck? Then I wonder whether I pack my colored crates because it makes me feel better about myself or because some prehistoric part of me has a desire to conform. Do I really know whether recycling helps the planet? Have I ever looked into it? What facts do I have? And if I don't really have any facts, why do I keep doing it?

Eventually, I couldn't stand it anymore. I needed to know whether recycling was really helpful or not, and just as I feared, my *beliefs* turned out to be way out of line with reality.

I was shocked to discover that, according to the Environmental Protection Agency, less than 3 percent of all the garbage generated by Americans is municipal waste. In other words, even if every person in every neighborhood in the country took personal empowerment seriously, the impact to the planet would be negligible.

I have the same kind of ambivalence about buying cheap goods.

With the explosion of megastores like Wal-Mart and Ikea, which specialize in foreign goods, many of us have become fearful that the U.S. economy is being overwhelmed with cheap Chinese goods. If we simply resisted buying these imports, we could apply economic pressure on China to reduce their carbon emissions. This would also strengthen our domestic economy, right?

But did you know we had similar worries in the 1980s when Americans panicked because the Japanese were flooding the U.S. market with cheap goods and also allegedly buying up all of Hawaii, Manhattan, and San Francisco?

Surprisingly, in both cases the facts don't support our paranoia. The impact individuals can have on restoring the trade balance is far less than imagined. I discovered that less than three out of every one hundred products sold in the United States today are from China. In the 1980s, the percentage of Japanese goods was also less than 3 percent. So even if every citizen in the country had boycotted Japanese products then, or Chinese products today, the net impact would be minimal—a lot like the net effect of recycling.

Later I discovered that less than 6 percent of all the merchandise in massive Ikea warehouses is made in China. Before I found this out, I would have bet money it was more than 50 percent.

Yet, even though I now have the *facts*, I still feel guilty when I buy a sweatshirt for ten dollars. I'm pretty sure if I'm not contributing to an

imbalance in trade and aiding polluters, then I am exploiting child labor somewhere or making it impossible for the store where I buy the
cheap sweatshirt to offer health insurance to its cashiers. Yet, on the
other hand, if I don't buy the sweatshirt, will the cashier be laid off?
Will some child laboring in a foreign country go hungry?

I constantly wonder if my personal choices are causing the dangerous problems we face today to worsen. I want desperately to do the
right thing, but I don't seem to be able to figure out what that is anymore. The problems and the solutions seem too multifaceted, too
complex. On virtually all fronts, I feel stuck. And also guilty.

This is what happens when individuals are held accountable for
complex social problems. The effect we can have on systemic problems becomes greatly exaggerated in an effort to transition responsibility from failing institutions, leaders, and experts to the man on the
street.

The Paradox of Pop Culture

To understand how pervasive the *personalization of blame* has become in the twenty-first century, just watch daytime talk shows for a
month. Popular television programs such as *Oprah, Dr. Phil, The
Suze Orman Show, Mad Money,* and *The Dr. Laura Show* all operate
on the same basic premise: We control our lives, and we suffer or
thrive as a consequence of our decisions. On a daily basis, celebrity
hosts offer quick tips on how to lose weight, save money, raise children, end poverty, have a successful career, improve a marriage, and
have a more meaningful relationship with God.

All we need is determination and a game plan.

Nonetheless, year after year the number of people who become
obese skyrockets. Bankruptcies and personal and national debt are
also rising. Divorces, victims of child abuse, alcoholism, and drug addiction are up, up, and up. Clinical depression, crime, pollution, high
school dropouts, and cancer—also all up.

With so much emphasis on self-empowerment, it doesn't take long
for viewers to start blaming themselves for their problems. They didn't

do enough; or didn't do it right: they weren't strong enough, persistent enough, smart or talented enough. By focusing almost exclusively on what individuals must do—and spending little or no time discussing the overwhelming systemic obstacles to real change—talk show hosts convince their viewers that their lives, families, neighborhoods, country, and planet remain troubled because of their own personal failings.

However, the vast majority of problems examined by daytime talk shows aren't really individual problems at all. They are rooted in highly enmeshed socioeconomic issues that have troubled humankind for many generations. So, the net effect of convincing individuals they can change their situation often winds up being as negligible as recycling or boycotting cheap foreign goods.

One of the best indicators that these problems are systemic, not personal, is the way television shows make their money. The topics for every program are carefully selected based on audience appeal—the larger the audience, the higher the price a network can charge for a thirty- and sixty-second advertisements. But when programming is driven by topics that plague millions of people, isn't that all the proof we need to realize a problem is systemic?

With such a large percentage of the population afflicted, these issues can no longer be considered individual problems.

The truth is, it has become so easy to mistake a systemic problem for a personal one that we are gradually abandoning systemic solutions that require huge investments and are slow to produce results.

Our views on obesity are a prime example of how this super-meme—*the personalization of blame*—inhibits real social progress.

The Obstinacy of Obesity

Thanks to experts everywhere, the vast majority of us now believe maintaining a healthy body weight is just a matter of personal discipline. We believe obesity is the consequence of making poor food choices and not exercising enough. So, similar to quitting smoking, a little bit of willpower is all it takes.

But is solving obesity really that simple?

If obesity were only a matter of personal willpower, then we should expect people who show great discipline in other areas of their lives not to succumb to weight gain. This would be inconsistent with the strength of character and determination they draw on to prevail over other challenges.

For a moment, however, consider one of the most willful, most powerful people in the world as evidence of why this *belief* is untrue.

For almost four decades Oprah Winfrey has hosted the most successful daytime talk show in the United States. Throughout her career she has also been a tireless cheerleader for personal accountability and change. Oprah has bravely tackled every topic from cosmetic surgery, racism, hormone replacement, autism, and infidelity to personal finances, breast cancer, rape, politics, and religion. She is responsible for a resurgence in reading through her book clubs and has raised millions of dollars for the homeless, for hurricane and tsunami relief, and for the victims of 9/11. Oprah has built orphanages, interviewed world leaders, introduced new consumer products, and launched an assortment of self-empowerment programs like *The Doctors, Dr. Phil,* and *Dr. Oz* to help individuals self-diagnose and self-correct.

If ever there were an icon for individual empowerment, it would have to be Oprah Winfrey.

Then in 2008, Oprah took a courageous stance on obesity. In a moving speech she publicly confessed to "falling off the wagon" again after successfully and unsuccessfully dieting since 1984. Winfrey admitted she had regained forty pounds in just a few short months.

Oprah Winfrey is not only educated, wealthy, and resourceful; she is both personally and publicly committed to personal accountability. And she is also able to surround herself with a staff of private chefs, personal trainers, nutritionists, counselors, and doctors—all dedicated to helping her maintain a healthy body weight.

So what does it mean when a stalwart like Oprah can't overcome obesity?

In Winfrey's case, true to form, she capitalized on her personal failing by transforming it into a week of programs designed to help her audience develop their "Best Self." She vowed to adopt a healthier lifestyle, eat right, exercise, and lose the gained weight. She courageously promised, yet again, to keep the weight off *this time.*

I would be a fool to bet against a powerhouse like Oprah Winfrey. That said, I will venture a guess that, without confronting the evolutionary and systemic forces that are the root cause of obesity, Oprah, try as she may, will be unable to keep her promise.

That's because obesity isn't a matter of willpower alone. It never was.

According to the Centers for Disease Control (CDC), more than one quarter of all Americans are now obese. Did these people all lose their willpower? And when it comes to children, obesity is spreading even faster. According to the CDC, "The occurrence of obesity for children between ages 6 and 11 more than doubled between 1980 and 2006, increasing from 6.5 percent to 17 percent." What's more, obesity-related medical expenses for adults and children accounted for over $47 billion in 1998, and the costs continue to rise.

The recent award-winning documentary *Super Size Me* delves deeper into the systemic nature of obesity: "America has now become the fattest nation in the world. Congratulations. Nearly 100 million Americans today are overweight or obese. That's more than 60 percent of all U.S. adults Since 1980, total obesity has doubled with two times as many overweight children and three times as many overweight adolescents." The film continues, "McDonald's feeds more than 46 million people around the globe each day, more than the population of Spain."

Within the last three decades obesity has become a dangerous worldwide problem. In fact in 1997, the World Health Organization declared obesity a "global epidemic" with over two billion overweight adults worldwide and with no relief in sight.

So here's my question: Isn't the term "global epidemic" all we need to recognize we are up against a larger opponent? As more of the world's population suffers the same health consequences as Americans, should we continue to blame and treat obesity as a matter of weak will? Or are there other powerful forces involved?

An Omnivorous Advantage

One of the most important yet rarely discussed causes of obesity is our evolutionary inheritance. As mentioned earlier, we aren't a blank slate when we are born. Every human being is born with instincts designed to help him or her prevail over environmental challenges.

During prehistoric times, the acquisition and judicious use of calories was vital to our survival. Our primitive ancestors developed three dispositions toward food that enabled them to thrive: (1) being attracted to the highest-calorie foods, (2) eating as much as possible when food was available, and (3) resting to conserve calories when they did not have to fight or look for more food.

In earlier times, no matter how much we ate and lay around, it would have been extremely difficult to achieve morbid obesity. Imagine how many bushels of apples a person would have had to find, pick, and eat to become overweight.

Today, however, with so many calorie-rich, processed foods available, all it takes is a couple fast food meals a day. It's hard to ignore the vast number of products in the grocery store that are laden with oil and fats and concentrated sugars such as high fructose corn syrup—or the fact these foods taste so much better than healthier choices. Oh, please. Do we really need to argue whether the rich flavor of real butter and cream or a fresh-baked cinnamon roll trumps a granola bar, banana, or rice cake?

Furthermore, we are a society that has made calorie-rich foods easier to purchase. They are arguably cheaper and more abundant, with a fast food joint around every other corner, a greasy food court in every shopping mall, and a snack concession in every movie theater, sports stadium, and amusement park. And we now devote an entire aisle in the grocery stores to different flavored chips and soft drinks.

Logically we may know we are consuming too many calories, but our biological inheritance conspires with massive social incentives to drive us toward fatty foods. We may know that we need to exercise, but we upgrade our televisions and watch movies and play video games instead. We may know that sitting and eating is harmful to our health, but we keep doing it anyway.

But how do we stop hardwired predispositions?

Obesity Is Systemic

I recently had the opportunity to meet Dr. John Ratey, professor of psychiatry at Harvard University and author of *Spark: The Revolutionary New Science of Exercise and the Brain.* Over the last decade Ratey has become an expert on the effects obesity has on the human body as well as the dangerous impact a sedentary lifestyle has on the brain's ability to learn and think. One afternoon we sat in his quaint Cambridge office, surrounded by paintings, books, stacks of research notes, and dog toys as Ratey relayed the alarming consequences of childhood obesity. His worry? *A society that holds adults accountable for their own obesity is a stone's throw away from making children accountable for theirs.*

In my view, nobody understands the magnitude of the systemic changes needed to stop the contagion of childhood obesity better than Ratey.

Ratey explained the details of a recent study conducted by the University of Florida. Scientists discovered that morbidly obese toddlers as young as four years old exhibited significantly "lower IQ scores, cognitive delays and brain lesions similar to those seen in Alzheimer's disease patients." On average, the obese toddlers trailed by twenty-eight IQ points, causing Ratey to conclude that overweight toddlers begin school with a mental handicap, destined for failure.

What disturbs Ratey the most are the most recent statistics from the Centers for Disease Control that reveal that one out of five four-year-olds in the United States is now obese. In other words, approximately 20 percent of all children entering the school system have a learning disadvantage. According to Ratey, these figures indicate a dangerous social trend that reaches far beyond poor food choices.

And being overweight doesn't just affect the cognitive abilities of children.

A second study examined senior citizens who averaged seventy-eight years of age. This research revealed that obese seniors had 8 percent less brain matter and that by simply requiring seniors to exercise forty-five minutes per day, cardiovascular fitness increased the volume of their brains and improved cognition.

Whether young or old, Ratey concludes, "too much fuel and not enough movement slows down the human brain."

The key appears to be an increasingly sedentary lifestyle that is harmful to cognitive processing. Every time the human organism contracts a muscle, it releases a protein that stimulates the brain to produce more cells and encourages brain cells to communicate with each other in new ways. In recent animal studies, adult rats that exercised increased both the protein and blood flow in their brains, causing them to demonstrate approximately 25 percent more intelligence than rats that remained sedentary.

Ratey's big insight?

The human organism may be a calorie rich–seeking organism, but it is also an organism that functions better when it moves. Over time, a sedentary lifestyle that is toxic to the human brain has become institutionalized, and we must do far more than blame children and senior citizens in order to fix the problem.

As a result of new research on the relationship between physical movement and cognition, Ratey has become a tireless advocate for requiring physical education and recesses in public schools, retirement homes, and offices. He campaigns for healthier, high-nutrition, and low-calorie cafeteria meals and the immediate eradication of fast food franchises on school grounds. He is a frequent traveler to the nation's capital and now consults with industry giants such as Google and Microsoft in a relentless effort to reintroduce movement into daily routine. According to Ratey, employees using "standing desks"—desks designed so that individuals can slowly walk on a treadmill while working at a chest-height work surface—are not only healthier, but they also make better decisions, remember more, and show higher productivity. During school, children who sit on large balls designed for movement, rather than in chairs, also generate cognitively beneficial proteins. And seniors who move around more generate new brain cells that stave off a plethora of mental illnesses.

The good news is that it doesn't take much to begin reversing cognitive impairments.

In a recent experiment, a group of senior citizens in New York were asked to participate in less than an hour of exercise a day, and the results were astonishing after just three months. The hippocampus, the area of the brain responsible for storing and coordinating memory, grew almost 30 percent in just ninety days. So, in addition to health benefits such as improved circulation, agility, stronger bones, and lower instances of depression, ADD, ADHD, and a growing list of psychological conditions, physical activity plays a vital role in the brain's ability to remember, learn, and solve problems.

Ratey distinguishes himself because not only has he made the important connection between obesity and cognition, but he is also one of the few experts working for massive, systemic change rather than holding the obese responsible for their own plight. He may believe in self-empowerment—the individual's power to rise above—but in parallel with personal accountability, he also acknowledges that powerful forces in society work against a healthy lifestyle.

Obesity isn't just a personal problem—it's a systemic one. And it needs a systemic prescription to cure it.

Perfecting Personal Responsibility

When B. F. Skinner published *Beyond Freedom and Dignity* in 1971, no one could have anticipated it would lead to assigning the blame for modern ills to individuals. Yet, personal accountability became one of the unintended consequences of the rise of behavioral psychology. In a short period of time, the entire world was introduced to the idea that *all* human behavior could be explained by a series of positive and negative reinforcements. This also meant that *all* human behavior could be modified by simply manipulating rewards and punishments.

Authority figures everywhere rejoiced.

Suddenly society had an easy, predictable formula for re-engineering human habits and inclinations. Rewards such as food, money, praise, and promotions could be used to strengthen desirable behavior. Conversely, punishments such as criticism, electric shock, isolation, and withdrawing privileges could be used to extinguish an array of undesirable tendencies. By consciously administering a regimen of "carrots and sticks," a person's behavior could be reshaped.

Overnight, teachers began plastering gold stars next to the names of high-achieving students in the classroom, middle managers got busy handing out "certificates of merit," and every player on the soccer team was singled out for their unique contribution and handed a trophy. Reinforce, reinforce, reinforce.

Aversion therapy clinics also popped up everywhere. If you wanted to quit smoking, lose weight, or stop raping and killing, all you had to do was receive a small shock of electricity every time you entertained the undesirable behavior. Light a cigarette, administer a shock. Light another one, administer another shock. Repeat this again and again and again until the brain finally associates cigarettes with pain and the individual no longer wants one.

In this way, natural temptations could be squelched, bad habits broken, and new behaviors learned.

Once a direct "cause and effect" relationship was established between undesired behavior and positive/negative reinforcement, it became an individual's "responsibility" to correct undesirable behaviors.

Suddenly, it was up to each of us to *reprogram* ourselves.

But, as the long-term results of aversion therapy disappointed (over time, people started smoking again, rapists began raping, and so forth), we came face to face with the true complexity of habit, free will, reason, morality, and inherited genetic predispositions.

It wasn't as simple as behavioral science once led us to believe.

But that didn't stop us from continuing to blame individuals for their circumstances. Rather than acknowledge the complex, systemic nature of human behavior, we merely substituted aversion therapy with self-help books, pharmaceuticals, and daytime talk shows and began beating the drum of self-empowerment louder and louder so as to drown out the voices of the few holdouts—the few psychologists—who were saying "not so fast."

As the doctrine of personal responsibility and accountability mounted, we lashed out at social victims, individual heads of corporations, experts, and world leaders, claiming they weren't doing enough. We also turned blame inward because we, too, weren't doing enough. Yet, even with all that accountability to go around, obesity, unemployment, illegal immigration, pollution, gang violence, drug addiction, and other stubborn problems continued to escalate.

Though the behavioral psychology movement may have launched a modern era of individual responsibility, in truth blame and self-empowerment are no match for the deeply enmeshed, systemic problems we must now address to assure our survival.

Business consultant David Gurteen, who calls himself a "Knowledge Management Facilitator," summarized the relationship between the understanding needed to solve complex problems and our tendency to blame when he said, "Where there is understanding, there is no blame."

Conversely, when understanding cannot be obtained, blame is the predictable way out.

As growing complexity makes the acquisition of knowledge increasingly difficult and causes global conditions to worsen, the personalization of blame will no doubt escalate, diverting focus away from defeating our most dangerous problems.

And while that may not sound like a hopeful prognosis for the future, *don't blame me.*

— 6 —

Counterfeit Correlation

The Third Supermeme

There is a popular story circulating about a physician who decided once and for all to explain to his colleagues what really causes heart disease. With a wry smile, he calmly relayed the following facts:

The Japanese eat very little fat and suffer fewer heart attacks than the British or Americans.

The French eat a lot of fat and also suffer fewer heart attacks than the British or Americans.

The Chinese drink very little red wine and suffer fewer heart attacks than the British or Americans.

The Italians drink excessive amounts of red wine and also suffer fewer heart attacks than the British or Americans.

The Germans drink a lot of beer and eat lots of sausages and fats and suffer fewer heart attacks than the British or Americans.

His conclusion: Eat and drink what you like. It's obvious that speaking English is what kills you!

THE THIRD SUPERMEME and obstacle to progress is *counterfeit correlation* or, as I sometimes call it, "Clavinism," after the know-it-all postman Cliff Clavin on the popular television program *Cheers*. Clavin was a character known for regularly citing trivia that he mistook for reliable research. He would always preface his statements with "It's a little known *fact* that . . ."

On the surface counterfeit correlation sounds a lot like Cliff Clavin or the physician explaining the cause of heart disease: It has the appearance of being firmly rooted in logic and empirical proof. But counterfeit correlation is a misleading form of logic. As a result, the conclusions drawn, though interesting, are false.

In the twenty-first century we have not only perfected counterfeit correlation, but we have fallen under its spell. Counterfeit correlation occurs as a result of three convenient practices:

- Accepting correlation as a substitute for causation,
- Using reverse-engineering to manipulate evidence, and
- Relying on consensus to determine basic facts.

Working together, these shortcuts have the power to make *beliefs* appear to be legitimate scientific facts. And in terms of progress, this spells trouble.

With so much information coming at us at such a fast pace, *counterfeit correlation* makes it impossible to separate facts from conjecture, opinion, theory, and authentic correlation. We simply don't have the time to check out every claim. Even if we did, the amount of information we would have to wade through to get to the empirical truth is overwhelming, and most of us don't have the skill or time.

But what's the big deal if we get a few facts wrong? It's not the end of the world.

Maybe not yet. But let's face it: Confusion over which are proven facts and which are unproven beliefs eventually has dire consequences. The Mayans, Khmer, and Romans found this out the hard way.

When we can't separate facts from fiction, we become highly susceptible to misdiagnosing our problems. This leads to pursuing one unsuccessful mitigation and remedy after another, all because we based our solutions on what appeared to be science, but wasn't. In this way, counterfeit correlation—passing off ill-conceived assumptions as fact—has become an extremely dangerous obstacle to progress in modern times.

The Canonization of Correlation

The reason counterfeit correlation has become so popular is easy to understand: Casually observing a relationship—any relationship—between two events is magnitudes easier than the grueling effort required to prove one thing actually *causes* another to occur. So as the world grows more complex and it becomes more difficult to isolate the root causes of our problems, we simply begin to *lower our standard for proof.*

It's a natural response to complexity.

But what exactly is correlation, and why is it inferior to proving cause and effect?

A correlation occurs when two things change at the same time. Nothing more. If the number of people who own handguns rises at the same time as global warming increases, then, according to counterfeit correlation, this *implies* the two events are somehow related.

But does this mean owning handguns *causes* global warming?

Hardly.

Or global warming *causes* handgun sales to increase?

Also unlikely.

Using an absurd example like handguns and global warming, simple reasoning is all we need to figure out the two incidents are not connected. There's simply no connection between climate change and handguns, and *it doesn't matter how strongly the data correlate.*

However, what happens when the relationship between two events isn't as clear? Take the relationship between drinking red wine and heart disease, for example, or vaccines and autism, subprime mortgages and a global recession, or teacher's salaries and public education.

Are these the same as handguns and climate change, or is there sound evidence one actually *causes* the other to occur?

That's the problem with correlation. Though it often *implies* causation, it doesn't necessarily mean that cause and effect have been *proven*.

In today's fast-paced world where the pressure to produce short-term results is omnipresent, the once-high standards for isolating cause have slowly begun to erode. Every day, experts draw new conclusions based on correlations. They hastily put out a press release claiming they have uncovered a long sought-after elixir to a massively complex problem and rush to reap the monetary rewards and fame that come with discovery. Then, without questioning their claims, the rest of society follows suit, prematurely implementing false remedies one after the other as time runs out.

When counterfeit correlations begin appearing everywhere—on the evening news, in a court of law, in classrooms, books, elections, and public policy—it becomes impossible to separate conjecture from fact. Remember what happened when Leinweber "discovered" that butter production in Bangladesh matched the rise and fall of the New York Stock Exchange? Counterfeit correlation. Or how about the famous 1954 McCarthy hearings that determined that anyone who had a Communist friend was a spy? Again, counterfeit correlation. Or the internment of thousands of American citizens of Japanese descent, stripped of their civil rights, during World War II because their race labeled them a threat to national security? All of these events were based on false correlations, erroneous cause-and-effect relationships that were quickly adopted as popular truths.

As if runaway complexity isn't enough of a challenge for the human brain, confusion over what are the essential facts amplifies our collective gridlock. If we can't accurately diagnose what is causing a problem, how can we fix it?

Still not convinced? Let me give a quick example that illustrates how difficult it has become to separate a fact from a belief and how this leads to chasing the wrong cures.

A few months ago, journalist Charlene Laino made the following report:

European researchers reported that teenagers who used their phones more than fifteen times per day were having more trouble falling asleep and staying asleep than were those who used their phones sparingly.

The revelation that cell phone use was disrupting the sleep of teenagers had parents everywhere telling their children to get off their phones or taking them away. Many schools reacted by limiting cell phone use, and psychiatrists, school counselors, celebrities, and trusted news anchors were urging the public to do the same.

But the report Laino and others filed was based largely on their interpretation of a study conducted by Dr. Gaby Badre, MD and PhD of Sahlgren's Academy in Gothenburg, Sweden, and the London Clinic in England. Dr. Badre discovered that in addition to high cell phone use, the teenagers who suffered from sleep disturbances also (1) drank more caffeinated drinks such as coffee and sodas; (2) consumed more alcohol; (3) had a general pattern of waking up later (11 A.M. instead of 8:30 A.M.) and (4) showed higher levels of anxiety and agitation. In other words, any one of these factors could also explain poor sleep. Badre's study never proved that high cell phone use was the *cause* of poor sleep, he only demonstrated that it, along with many other factors, was *correlated* with it.

It was a clear case of Clavinism at work.

But when credible sources, like those in the media, tell us that sleep disturbance has been associated with cell phone use, what conclusions do we draw? We immediately assume we need to grab control of our teenager's cell phone. Ironically, this is likely to produce the exact opposite effect: Badre's research, along with another study conducted by a former professor at Rutgers, Sergio Chaparro, suggest that a teenager can experience *more* anxiety when one of the venues for *alleviating* stress—such as talking on the telephone—is eliminated.

So what should a parent to do? Take the phone away, or encourage them to use it more?

In this way, counterfeit correlation makes it extremely difficult to separate fact from fiction, cause and effect from mere correlation.

This, in turn, makes it confusing and challenging for individuals, families, schools, leaders, and nations to act responsibly.

Still, we are a society that is extremely quick to accept correlations as proof of fact.

At one time we were convinced Hormone Replacement Therapy reduced the risk of heart disease. I, along with other women across the country, flocked to the doctor for a prescription. Then, less than a year later a different researcher proved HRT actually *increased* heart disease. One year later, physicians rejected the study and *returned* to recommending HRT again. We played this same game of ping-pong when boxer shorts were shown to be better for fertility than briefs, then weren't; when driving a hybrid car was believed to be better for the environment, then wasn't; when Saddam Hussein was considered the greatest threat to America, then wasn't; and when we thought the flood of cheap Chinese goods was causing a dangerous trade imbalance, and later, we learned better.

It's safe to say we have entered a period when our standards for proving cause and effect have become dangerously low. We are willing to accept simple correlations in lieu of spending the time and resources to understand highly complex and often threatening issues. In turn, this has led to a plethora of failed programs that all claimed to be the cure to our most difficult challenges.

Recently I came across a Web site maintained by Dr. Jon Mueller, a professor of social psychology at North Central College and expert at discerning the difference between causation and correlation. Mueller's site offers the following news headlines as blatant examples of how easy it is to confuse correlation with cause:

"Fan in Room Seems to Cut Infant's Risk of Crib Death"
"Texting Improves Language Skills"
"Study Suggests Attending Religious Services Sharply Cuts Risk of Death"
"Breast Implants Lower Cancer Risk but Boost Suicides"
"Migraine Often Associated with Psychiatric Disorders"
"Background TV Harms Tots' Attention"

"Some Types of Cancer Raise Divorce Risks"
"Social Isolation May Have a Negative Effect on Intellectual
  Abilities"
"Keeping a Food Diary Doubles Weight Loss"
"Eating Fatty Fish Lowers Risk of Dementia"
"A Surprising Secret to Long Life: Stay in School"
"Fed Says Fear of Hell Makes Us Richer"
"Disciplinarian Parents Have Fat Kids"
"Sexual Lyrics Prompt Teens to Have Sex"

And the list goes on. Once we start looking for them, inaccurate correlations are everywhere.

In fact, the confusion over correlation and causation has become so commonplace that psychologist and satirist, Shane T. Mueller, recently described a Gallup Poll where 1,009 Americans were asked, "Do you believe correlation implies causation?" Surprisingly 62 percent replied "yes!" Mueller concludes: "It's really a mandate from the people. It says that the American people are sick and tired of the scientific mumbo-jumbo that they keep trying to shove down out throats, and want some clear rules about what to believe. Now that correlation implies causation, not only is everything easier to understand, it also shows that even science must answer to the will of John and Jane Q. Public."

Science "must answer" to the will of the public?

Aren't facts, facts?

Mueller brilliantly exposes what many of us have suspected all along: counterfeit correlation has infiltrated the once-sacred fortress of rational thinking where facts were held up to scrutiny.

Perplexing Public Policy

What happens to public policy when economic, political, and social problems are misdiagnosed as a result of false correlations?

Earlier in this book I talked about the seasonal water shortages in Northern California that have made it difficult to irrigate vast agricultural lands and provide sufficient drinking water to an exploding population. Every year, as the rains decrease, the situation in California grows a little worse. But sadly, drought is an undeniable historical fact in this region of the country.

And so is housing development.

In 2002, California reported having almost thirty-five million residents—up five million in ten years. And by 2030 the state projects the population to reach fifty million. As, by some estimates, the seventh largest economy in the world and a hotbed of tomorrow's technology, the area shows no signs of slowing down.

One of the unspoken ways the small town I live in controls development is by limiting the number of showers, toilets, and sinks each home is allowed. Each home is assigned "water credits," and every water-using appliance is assigned a specific number of credits.

Although on the surface this may sound bizarre, on some level it seems to work. Long ago, tiny beach cottages would have been torn down in favor of giant beachfront mansions if not for the fact that an eight-thousand-square-foot home with one bathroom doesn't make a lot of sense. Water restrictions, along with an enthusiastic Carmel Heritage Society, have kept the town of Carmel small, quaint, and a worthy destination for thousands of tourists who flock each year to admire the storybook homes and shops.

Recently, I undertook a small remodeling job on a small two-story cottage that never had a bathroom on the first floor. I decided to convert a coat closet into a powder room so guests would no longer have to go into someone's bedroom upstairs to use the facilities. Since I had a few "water credits" still available on my property, I went to the county office, paid my fees, and applied for a permit to install one toilet and one sink.

My application was quickly denied.

When I inquired about why it had been denied, I received a lecture from an aggressive clerk about the water problems in our area. After fifteen minutes of scolding, to my relief she finally concluded, "We can't just keep letting people add a toilet every time they feel like it! What do you think would happen?"

I thought about this for a moment. Then I replied, "But having more toilets doesn't make me go to the bathroom more. I flush the same number of times whether there is one or fourteen toilets in the house."

She looked confused. Then she shrugged and denied my permit anyway.

It was obvious to me the number of toilets had been incorrectly correlated with water use, but apparently this minor detail doesn't make any difference in determining public policy anymore.

The fact is that flushing only increases when the number of occupants in a house increases, not when the number of toilets does. What was unclear—unstudied—was the relationship between the number of toilets and the number of occupants in a household. This was the critical missing data on which public policy *should* have been based.

But that's just one small example of how counterfeit correlations conspire to produce irrational policy. There are larger, more serious consequences of this supermeme as well.

When we discovered a relationship between AIDS and homosexuality, gay men became ostracized and, as fast as lightning, we labeled the disease a "gay problem." When we determined there was a relationship between nutrition and crime, a program to change the diet of prisoners was heralded as a new crime-prevention tool. And when we correlated smoking with cancer, we moved to ban smoking everywhere.

Then, as time passed we determined AIDS wasn't unique to the gay population. It's a virus that is transmitted through bodily fluids and doesn't discriminate by sexual preference. And though diet is important, it did little to reduce prison violence or a growing crime wave. Then came news that it was not the cigarettes themselves that *caused* cancer, but rather the harmful chemicals added to tobacco by the manufacturers.

We also learned much later that children who sleep with the light on are not more likely to develop myopia, but they are likely to have myopic parents who leave a light on in their child's bedroom because they themselves need the extra light and who pass on the genetically inherited trait to their children. We also changed our positions on flu vaccines, NAFTA, racial profiling, and the practice of waterboarding all because false correlations were mistaken for cause.

Although quick correlations may be attractive, they do little to unravel the complex, interlaced causes of the problems humankind must now solve.

The more we rely on false correlation and lower our standards for proof, the more we open the floodgates for irrational beliefs and behaviors. This has been the historic pattern of all great civilizations, and we in the twenty-first century are not immune.

Take the problem of a deteriorating public education system as an example. There is so much counterfeit correlation—opinions and correlated studies—that attempt to explain why our public schools are failing that it has become impossible to sort out what the real problem is anymore.

It is well known that public education in the United States has been consistently falling behind global standards for several decades now. New standardized testing shows a dangerous decline in mathematics, science, and writing scores across all age groups. Although the overall high school dropout rate has declined from 15 percent in 1972 to 11 percent in 2008, this progress has regrettably occurred along racial and socioeconomic lines. Today, only 2.7 percent of children from high-income households drop out of high school, whereas almost 24 percent from low-income households withdraw early. A more careful examination reveals that almost one-third of the Hispanic students drop out, followed by 14 percent African American, compared to only a 9 percent dropout rate among children of Caucasian descent.

One of the unhealthy consequences of the high dropout rate, along with increased parental pressure on children from high-income families to attend a top university, has been "grade inflation." Grade inflation is a trend whereby students receive higher grades for performance that would have otherwise earned them a lower grade. Since the early 1960s grades have been steadily creeping upward. Depending on which study you trust, the increase ranges from one-quarter to one full grade in the past ten years, all the while standardized tests show that actual learning is headed in the opposite direction.

There are a lot of benefits to giving out high grades. Less homework, fewer and easier tests, better report cards, and higher college admissions make everyone happy—the students, the parents, the school, the school district, and the State Superintendent of Schools.

Unfortunately, there is one small side effect to this love fest: No one is learning anything.

There has been an ongoing, passionate debate for many decades about how to repair the declining public education system. Strangely, race and socioeconomic issues rarely enter into the discussion anymore despite evidence that the dropout rate is divided along racial and economic lines. Presumably, when you fix the entire system, every student from every racial background will benefit. At least that's the assumption.

One group argues that teachers need to be made more accountable, and this group suggests the abolition of tenure and instituting regular tests for teachers. Some argue teacher salaries are too low and that this leads to hiring inferior educators unsuitable for other higher paying jobs in society. Others blame unions for keeping unfit teachers in the system long after their effectiveness has expired.

Still another group argues that the absence of parental involvement is the reason. In two-income households, many parents don't know what their children are studying at school, whether they are having problems, or even if they have homework. Other people blame parents for "overbenefiting" children and imbuing them with a sense of entitlement wherein kids are no longer motivated to achieve. Still others blame the government for underfunding education. They claim the root "cause" of teacher and student failure is the lack of money for teacher training, dilapidated facilities, outdated textbooks and computers, unhealthy meals, and low teacher pay.

But these are only some of the reasons public education may be failing in the United States.

When we add to these arguments the growing number of children who are now depressed, suicidal, or homicidal, a clearer, more disturbing picture emerges. The problem with public education is a highly complex, systemic one that is not likely to be resolved by simply updating textbooks and raising salaries.

Dr. Yaneer Bar-Yam summarizes the dangerous role false correlations play in incorrectly diagnosing the failure of public schools:

> In 1983, The National Committee on Excellence in Education issued a report entitled *A Nation at Risk* which was a call to arms to improve the [U.S.] education system in the face of international competition. According to the results of standard math and science scores the U.S. is far behind many countries in the world today. The results, if anything, have become worse over time rather than better. If we look at test scores as an indication of success in the future, this seems to be a paradox: we have had many years of low scores and still the economy is remarkably strong—how should we interpret this data? Does this mean math and science have nothing to do with economic success, or that it has an inverse relationship to success? Surely this seems unreasonable.

Bar-Yam makes a compelling argument here. An inverse relationship between declining math and science scores and growing economic potency might lead us to conclude that a less-educated population is more productive and more economically viable. However, consider for just a moment the implications of coming to such a conclusion. Imagine the impact that correlating the wrong statistics in the wrong way would have on the future of public education. For example, if the economy grows as math and science scores decline, why not eliminate these subjects altogether?

Understanding causal relationships under chaotic, complex conditions requires a commitment to the highest standards of scientific proof no matter how long it takes. Yet, in an impatient, *correlation-addicted world*, scientific standards have been lowered to accommodate simplistic correlations that often have little or no relevance to root causes.

Correlation and Gridlock

When a society accepts correlation as a quick substitute for causation, it launches social programs aimed at a million symptoms without ever getting to the root of its problems. For a brief period of time the symptoms may ameliorate. The problem may even seem to disappear. However, later the problem reemerges, but this time with superpowers.

Short-term amelioration is a ruse—and a dangerous one.

The confusion over water conservation is a perfect example of how false correlations cause the impression that water shortages can be fixed by simply *using less water*.

Dr. Martin Parry, professor at Grantham Institute for Climate Change and Centre for Environmental Policy at the Imperial College in London, was the first person to shed light on the role that mitigations such as conservation play in solving complex problems. In his view the purpose of mitigation is merely to lessen the severity a society must, in the end, face to survive. According to Parry, insufficient mitigations lead to "adaptation deficits" which eventually add up to extreme and painful corrections.

Parry understands that although successful mitigation can temporarily reduce the size of a problem, *it can never solve it*. That's why we call it a "mitigation." As cochair of the 2008 Inter-Governmental Panel on Climate Control, Parry applies his observation to the ecology movement: "Mitigation has got all the attention, but we cannot mitigate out of this problem. We now have a choice between a future with a damaged world or a severely damaged world." No matter what the problem is—whether it's a cataclysmic economic, political, or social issue—the drastic measures we will have to take when a problem threatens our survival will depend on what things we did along the way to *lessen* the impact of that problem.

But the key here is "lessen" the impact of the problem. Mitigations are not designed to cure our problems. According to Parry, a boulder may be rolling downhill headed straight for us, but the extent to which we slow it down and chisel pieces off of it determines how much damage we will suffer when the boulder hits the bottom.

Reverse Engineering

Replacing correlation with causation is one thing, but there are also other ways in which facts become obfuscated when a society reaches the cognitive threshold and begins accepting beliefs in lieu of knowledge.

In the twenty-first century we have become experts at amassing facts to support the conclusions we want, and then organizing this data into logical, plausible arguments. In marketing we call this "spin," but in the scientific community this is known as "reverse engineering."

For example, this is how we determine the cause of air disasters. Following an air disaster, it is customary to recover the "black box" cockpit recorder and reassemble all of the remaining pieces of the aircraft to determine what caused the plane to fail. We use this same technique to pinpoint arson fires, fix automobiles, and, in science, to "back into" unobservable phenomena like quarks.

The original intent of reverse engineering was to deconstruct an event into smaller principles that could be replicated in order to fully understand what had occurred. The process requires starting with the end result and then working backward until all the contributing pieces and processes have been accounted for.

To a large extent this is the same process we use to reach conclusions about the black stars and remote planets that are too far away to study firsthand. We acknowledge a new phenomenon and then comb through our grab bag of science to retrofit laws and theories known to explain the reaction. Then we create theoretical tests to see which explanations stand up. This is how new theories become accepted over time.

Because reverse engineering is so useful for explaining things millions of miles away, we also now use it as a convenient way to explain things closer to home. However, we often go a little further than mere explanation. We manipulate the facts into what is now commonly referred to as "spin." In fact, in recent times reverse engineering has become one of the most powerful tools in mass marketing; a series of facts can be strung together to produce almost any desired result. As the old saying goes, "If the facts don't fit the theory, just change the facts." Is it any wonder we have become confused about which are empirically proven facts and which are unproven beliefs?

I learned an important lesson about how easy it is to retrofit a story to fit the facts a few years ago when I was still an executive in Silicon Valley.

I was working for a client who had developed a new cellular telephone but who also had managed to miss the market. Instead of smaller phones with miniature keypads and color touch screens, the client had developed a large portable phone with a large keypad and equally cumbersome battery. It was awkward looking, unattractive, and clunky—not fit for any pocket.

Using reverse engineering, our marketing firm tried to imagine a use for this phone. We conducted research nationwide, including live focus group studies, surveys, and more. When all the due diligence was done, we had our answer: The large phone with limited functions was ideal for senior citizens. The phone was repackaged and became an immediate hit among seniors who needed a larger keyboard and only desired the ability to make and receive calls. So the company was immediately repositioned as a provider of user-friendlier technology for seniors. All their literature, Web site, and press materials were retrofitted to support this new brand position: easy-to-use, larger, streamlined technology for seniors.

Job well done.

Recently, while attending a local charity function, I had a chance meeting with the new CEO of the company. He did not know I had previously worked with the company, so he began telling me all about the company's successful history, explaining that it was specifically founded to provide technology for a growing senior market. He was sincere in his description of the firm's origins, but he was also painfully inaccurate. In his view, the success of the senior cell phone was the deliberate, planned outcome of a premeditated corporate strategy. But in truth that strategy had been retrofitted to turn a product failure into a success.

Clearly, revisionist history had taken place.

In this way, reverse engineering has become an accepted practice in modern society. Facts become obscured, and history is simply revised to fit new conclusions—making knowledge all that much more difficult to discern.

In politics, reverse engineering has also become commonplace. Frequently, when presidents retire, their true intentions are often reverse engineered to fit their public record. Today, President Nixon's success in opening relations with China makes his involvement in the Watergate scandal look like a partisan vendetta. Even the invasion of Iraq, originally based on the threat of "weapons of mass destruction," has been repackaged as a unique opportunity to bring democracy and stabilization to the Middle East. Take any leader from any nation and watch how, over time, their errors are retrofitted into a logical, heroic, idealized story. But this should come as no surprise. When complexity makes facts and knowledge difficult to attain, we have established that we are an organism that is susceptible to beliefs. Fictionalizing history is just one more example of this phenomenon.

When reverse engineering is applied to highly complex problems such as a global recession, terrorism, or air safety, it also produces inaccurate and dangerous conclusions. For example, in his treatise "Reverse Engineering Gone Wrong: A Case Study," A. J. McEvily illustrates how complexity does not lend itself to deconstruction:

> At the time of the 2000-hour overhaul, the [private aircraft's] shaft and bushings are inspected, and if overly worn they are replaced. In the present case, it was found that the shaft and bushings were in need of replacement. However, the original manufacturer of the butterfly valve had gone out of business, and a second manufacturer supplied the replacement parts.
>
> The second manufacturer used a process known as *reverse engineering* in manufacturing the parts. That is, he attempted to copy in detail the characteristics of the original shaft and bushings. However, the importance of proper heat treatment was overlooked, and as a result the butterfly valve malfunctioned in flight and the plane crashed.

The attempt to duplicate a simple butterfly valve reveals how easily essential characteristics can be overlooked and incorrect conclusions drawn as we attempt to deconstruct many, many interrelated processes. The probability of missing just one ingredient grows exponentially as the complexity of our technology and circumstances mounts. So when it comes to air safety, financial bailouts, escalating crime, and the disposal of nuclear waste, reverse engineering often leads to catastrophic results.

The Democratization of Fact

Late one evening I listened to popular comedian Jon Stewart bemoaning the popularity of Wikipedia on the Internet.

In case you're not familiar with Wikipedia, it's a collaborative, free, online encyclopedia that can be edited, added to, and improved by anyone regardless of their credentials, background, or bias. The idea behind the site is that errors will be organically corrected by the users of Wikipedia themselves: If they read something inaccurate, they have an opportunity to correct it in real time. In this way, the founders of Wikipedia hope the cumulative information of humankind stays current, accurate, and accessible to the general public.

To give you some idea of how successful Wikipedia has become, in 2009 the site offered more than three million articles on a broad range of subjects and the online encyclopedia was growing fast. According to the founders of Wikipedia, "In a *past* comparison of encyclopedias, Wikipedia had about 1,400,000 articles with 340 million words in total, and the Encyclopedia Britannica had about 85,000 articles with 55 million words in total, and Microsoft's Encarta had about 63,000 articles and 40 million words in total."

What's more, according to comScore, an Internet user-measuring and -monitoring service, approximately 326 million users a month now use Wikipedia as a source for information.

In response, Jon Stewart recently reminded viewers of the dangers of relying on anonymous experts, particularly when there is no assurance that information is correct: "Since when are facts something you vote on? According to Wikipedia, if we all agree, then it's a fact. Is that all it takes? Just agree? I thought facts were something else."

What does it mean when facts are determined by consensus? And is this unusual? After all, polls indicate an overwhelming majority of Americans believe O. J. Simpson murdered his wife despite being exonerated by a jury in a court of law. Isn't that fact-by-consensus? How about blaming three auto executives for the failure of the American auto industry? Also fact-by-consensus. And let's not forget the benefits of recycling our garbage every week. Again, unilateral consensus.

As we now know, when facts become extremely difficult to acquire, irrational beliefs take their place. Sometimes these beliefs are based on false correlations, sometimes they are the result of manipulating logic, and sometimes they are simply based on popularity.

But without a firm grip on the facts, we lose our ability to determine what is causing our biggest threats. So we simply begin accepting beliefs in lieu of knowledge.

One afternoon, I had the opportunity to watch the former Chairman of the Federal Reserve and financial icon Alan Greenspan make one of his usual twice yearly reports to members of Congress. As Greenspan carefully parsed his words, making certain not to stir up the oppositional and personalization of blame supermemes, he calmly addressed one implied correlation after another. Finally, in a rare moment of exasperation he looked up at the panel and reminded them, "Gentlemen, you are each entitled to your opinions, but you are not each entitled to your own facts."

An apt warning from the grand master of complexity himself.

— 7 —

Silo Thinking

The Fourth Supermeme

On September 9, 2009, two legends in science, James Watson—who along with Francis Crick won the Nobel Prize for discovering the structure of DNA—and E. O. Wilson, world famous naturalist and father of sociobiology, came together on the stage of an intimate theater at Harvard University. Wilson, now in his eighties and dressed in his customary loose sports coat, khakis, and tie, had the lean, spry look of a graduate student. Watson, also in his eighties, arrived in a white linen suit with all the authority of a Senator from the South. As the two titans sat down on opposite sides of the moderator, the room grew quiet with anticipation.

It was a historic moment. Watson and Wilson had been fierce rivals in the world of biology for almost a quarter of a century.

Sometime in the 1950s Watson had the nerve to compare Wilson's work in natural history to "collecting stamps." And this got the ball rolling.

Then Wilson countered, suggesting that Watson's research might be a better fit for the chemistry department because he clearly misunderstood the purpose of biology.

Over time, as molecular biology garnered more and more attention, the antagonism between the two scientists grew more public.

Watson insisted that more budget and resources should go toward micro-biology research. He dismissed evolutionary biology as old hat, denunciated ecology, and declared that genetics was biology's legitimate frontier. Wilson then punched back: In his autobiography, Wilson called Watson "the Caligula of biology."

Things went back and forth this way for several decades. Biology was frac-tured down the middle, with two giants battling for its future.

But on this particular evening, in a dramatic turn of events, both men seemed sincere in recognizing each other's contributions to science. Wilson was the first to acknowledge that the acrimonious competition had, in the end, served both men well. Watson had been such a worthy opponent that it spurred Wilson to step up his own research and approach. When confronted by the moderator about his early resistance to molecular biology, Wilson made an astonishing admission: "I was wrong," he shrugged. He then described how crucial Watson's work turned out to be to evolutionary biology: Thanks to the discovery of DNA, biologists finally had the tools they needed to trace the evo-lution of all species. Much of the guesswork had been eliminated.

Watson agreed with Wilson. In his view there was nothing to argue about anymore: Evolution is explained by cellular biology, and cellular biology is clearly explained by evolution. Over time, the two had converged.

So, Wilson and Watson had called a truce.

Whether it was the wisdom and grace that comes from age or the natural progression of science or Wilson's belief in the unification of knowledge, which came in 1998 with his publication of *Consilience,* a once-bitter rivalry was re-placed by a deep and abiding respect. That night, two giants in biology came together on a small stage at Harvard, old friends joined by the common desire to unravel the secrets of life on earth.

IT'S OFTEN THE CASE that the big picture is impossible to see, espe-cially when we keep holding the smaller one in front of our face. Per-haps at first we can't understand what the double helix of DNA has to do with evolution and vice versa, or maybe we wonder how religious doctrine can ever be reconciled with the principles of evolution. Perhaps at first blush it's unclear what physics has to do with psychology, psychology with geology, and geology with economics.

But so what?

As Wilson and Watson discovered after decades of feuding, *that doesn't mean one vantage point is right and the other is wrong.* It simply means we haven't discovered how they work together in the context of a larger, more expansive system. That's sometimes due to our own myopia, but it's also often due to the fact that our brains may not yet be capable of making those complex connections. After all, there are biological, cognitive limits to what we can understand. Sometimes the big picture is just too big.

Impenetrable Silos

There is a natural tendency in the human brain to try to reduce complexity into discrete, manageable components. This phenomenon manifests itself in many forms: impenetrable corporate divisions, government departments, academic disciplines, and religious sects. Generalists are replaced by "specialists" in increasingly narrow fields, and strategic objectives are sliced into bite-sized targets that can be measured and tracked and for which individuals can be held accountable. If the sum is greater than the parts, then why not just start with the sum and break it down into smaller parts to begin with? Sounds like a reasonable approach.

But is it?

The fourth supermeme is silo thinking: compartmentalized thinking and behaviors that prohibit the collaboration needed to address highly complex problems.

Instead of encouraging cooperation between individuals and groups that share a common objective, silo thinking causes undermining, competition, and divisiveness. As silos prevent sharing and coordination across organizational boundaries, information that is already difficult to acquire becomes even more inaccessible.

Carol Kinsey Goman describes the effect of silo thinking in her article "Tearing Down Business 'Silos'":

> I've seen firsthand what silos can do to an enterprise: The organization disintegrates into a group of isolated camps, with little incentive to collaborate, share information, or team up to pursue critical outcomes. Various groups develop impervious boundaries, neutralizing the effectiveness of people who have to interact across them. Local leaders focus on serving their individual agendas—often at the expense of the goals of the rest of the organization. The resulting internal battles over authority, finances and resources destroy productivity, and jeopardize the achievement of corporate objectives.

Today silos exist everywhere. The CIA doesn't speak to the FBI and the physics department doesn't set foot in the economics building. Environmentalists don't talk to oil executives, defendants don't talk to prosecutors, Republicans don't talk to Democrats, doctors don't talk to insurance companies, and Al Qaeda doesn't talk to anyone.

And we wonder why society is gridlocked and broad, complex, systemic problems continue to worsen.

The Three Kluges

One of the most frightening examples silo thinking has on arresting human progress can be found in the health care industry.

According to the Institute of Medicine, from 1993 to 2003 the population in the United States increased 12 percent and emergency visits jumped over 27 percent. And recent surveys indicate the problem is growing exponentially worse every year.

But why?

On a recent cross-country flight I had the good fortune to sit next to the executive vice president of one of the nation's largest health insurance companies. He was a well-dressed, articulate former surgeon who sold his practice many years ago for the purpose of bringing real change to health care. From his perspective, insurance companies were holding most of the cards.

When I asked him about what he thought about the U.S. "public option" health care initiative, he said that *no increase in Medicare or Medicaid expenditures would resolve the problem.* He then explained that the best way to understand the health care industry is to think of it as three independent silos: hospitals, doctors, and insurance companies.

He went on to explain that today, when we are admitted to an emergency room, the ER physician who must treat us has almost no information regarding our health prior to our admission. When we are rushed into emergency, we are a total stranger to the attending doctor.

If we are lucky, someone who has accompanied us tells the doctor what they know about our medical history: what medications we take, who our regular doctor is, what we are allergic to, and any other information that comes to mind at the moment. Sometimes the ER staff can obtain access to some of our medical records and stitch together a profile that can help the doctor make a diagnosis, but it's a random, hurried, and flawed process at best. More often than not, the attending physician has to take his best guess by interpreting our symptoms on the spot, without the benefit of knowing any details about our medical history.

The ER doctor we are depending on at this critical moment has no baseline information about us, so things that may be perfectly normal for *our* body, but not for anyone else, are very likely treated as one of the symptoms related to our condition. For example, the attending doctor has no way of knowing that our heart has always run a little fast since the day we were born or that our temperature has always been on the high side or that we've had puffy ankles all our lives just like our mother, grandmother, and great-grandmother.

Instead, the ER physician is trained to use the parameters for what is "average." Any symptom that falls outside of this average is interpreted as an "irregular" symptom that gets factored into the final diagnosis—no matter whether that symptom is perfectly normal for our body or not.

So for the most part, ER diagnoses and treatments are speculative—the attending physician's "best guess."

And the statistics that result from taking "best guesses" are sobering. According to the *New England Journal of Medicine* there is a 34 percent probability that an ER patient will be readmitted within 90 days.

But here's the real shocker.

Within one year over 50 percent of those who were discharged after a surgical procedure were re-hospitalized or died within 12 months of their discharge.

So what kind of emergencies are these patients being treated for?

When the highest rates of readmission within the first thirty days were examined, it revealed the following diseases were the most likely candidates:

1. Congestive Heart Failure, 27 percent
2. Psychoses, 25 percent
3. Vascular Surgery, 24 percent
4. Chronic Obstructive Pulmonary, 23 percent

A quarter of the patients who suffer heart failure, psychosis, vascular surgery, and pulmonary problems return to the ER in four weeks. Could that be possible?

In theory, any patient with a serious ailment would see their regular doctor following a visit to the emergency room so they wouldn't have to return to the ER.

This is where we move into the second silo: doctors.

Today, doctors treating Medicare patients have to see four to six patients an hour in order to stay in business. Do the math: This means any doctor who accepts Medicare today can spend ten to fifteen minutes per patient before he starts paying out of his own pocket.

This means that when we have a serious complication following our first ER admission, our doctor, who has already used up his fifteen minutes, has two choices: If there's time, he can send us to a specialist who is also allotted another fifteen minutes to see us; otherwise, he directs us straight back to the ER—the same place that misdiagnosed us the first time around. In a growing number of cases, patients are simply instructed to return to the ER.

Then things start going downhill fast.

While our doctor (silo 1) may not be able to afford seeing us as often as he needs to, the hospital (silo 2) is happy to readmit us to the emergency room as many times as we need. The hospital makes the same amount of money whether we are a new patient or a returning one. Actually, returning patients now represent a significant percentage of a hospital's revenue stream, so in one respect, misdiagnosis followed by readmission has become an unexpected source of income. In 2009, over $17.4 billion was paid to hospitals for emergency room readmissions. This number is exponentially climbing as doctors direct seriously ill patients—who can't be helped in the allotted ten to fifteen minutes—back to the ER for help.

And then there's the third silo: insurance companies.

Every time I am admitted to the ER, my insurance company pays top dollar to a hospital that makes money from readmissions. My doctor knows the ER will gladly accept me, so there is no incentive for him to do anything differently. And insurance companies just keep passing along higher premiums to cover their expenses. The ER doesn't have the medical histories they need to diagnose and accurately treat critically ill patients, so the number of deaths and readmissions continues to skyrocket out of control, as does the cost.

It's a vicious cycle, and all the while the patients pay the price.

In 2009, Anthem Blue Cross of California raised my health insurance premium almost 30 percent, despite my having a clean bill of health. When I called to ask how they could justify the steep increase, I was told, "This happened across the board on all policies, not just yours. We have to raise everyone to cover the additional costs. It isn't personal; everyone got the same raise."

Not personal? Then why is the money coming out of my *personal* checking account?

Then, in February 2010, President Obama mentioned Anthem Blue Cross in a speech to the nation regarding health care reform. He singled them out as one example of insurance premiums gone amuck. Shortly thereafter, the U.S. Congress and California Insurance Commissioner opened their own investigations into Anthem's premium hikes.

But pinpointing Anthem Blue Cross looks a lot like blaming a few auto executives in Detroit or a handful of financial institutions for the global recession. Holding one insurance company accountable isn't going to ameliorate the silo thinking that is the root cause of the health care crises. The next thing we know, we will be asking Anthem executives whether they flew into Washington on private planes.

Today, the United States pays the highest per capita for health care of any country in the world, but in terms of quality, it ranks thirty-seventh among 191 member nations of the World Health Organization.

To put it mildly, there's a lot of room for improvement.

The executive I met on the plane saw a colossal opportunity for private insurance companies to reduce costs and improve services by simply making "baseline" patient information and medical histories available for every patient admitted into the ER—something that would require all three silos to work together. This would immediately improve ER diagnoses and decrease the number of readmissions, which in his view was one immediate way to stop escalating costs.

Similar to Dean Kamen and other innovators, he also agrees that the biggest challenge the U.S. faces isn't technology, but rather silo thinking. Even though everyone seems to know what the problem is and how to fix it, hospitals, doctors, and insurance companies believe that collecting, updating, and distributing patient information is someone else's job.

Silo thinking makes it extremely difficult to fix a systemic problem because no one silo is responsible for the overall problem.

Health care is just one example of silo thinking. Silos stand in the way of solving systemic problems everywhere we look.

Until the recent attacks of 9/11 and the 2009 failed attempt of the Detroit-bound Northwest Airlines flight, we were content to allow the CIA and FBI to each run their own show with very little information being shared between the two agencies. Then new attacks suddenly heightened the need for greater cooperation between government agencies. There was also an acknowledgment that sharing more information between the government and academia—information that was extremely difficult and costly to obtain—would be beneficial to the research departments of major universities. In keeping with this, the CIA reinvigorated the MEDEA program (Measurements of Earth Data for Environmental Analysis) to make previously classified information available for environmental studies. This meant that costly information that the CIA already possessed would not have to be duplicated by higher learning.

It didn't take long before the CIA came under fire for making an attempt to collaborate. In *Business Week*, Saul Kaplan reports that "talking heads across cable news accused the CIA of negligence, arguing that sharing data with environmental scientists was a distraction from its core mission of minding the American public." Kaplan then went on to chastise the silo mentality:

> But the pundits have it wrong. The CIA and all Homeland Security organizations should be doing more, not less, cross-agency collaboration and data sharing. The protection of data, capabilities, and turf has gotten us into the current mess. Perhaps if the focus had been on networking capabilities and sharing data across silos, America would be a safer country today.

So, as *silo thinking* becomes an entrenched supermeme, it becomes increasingly difficult to pool resources, and as a result information becomes even more difficult to acquire. Time and resources are wasted as each agency attempts to independently duplicate the work of others and the collaborative solutions needed to solve our most complex and dangerous issues grow more distant.

Another example of how silo thinking inhibits progress can be found among the growing number of nonprofit organizations. It is shocking to discover how many independent nonprofit organizations today are out to solve the same problems. In Monterey County where

I live, there are approximately 1,212 registered nonprofits. Translated, this means there are roughly three nonprofit organizations for every 1,000 residents—each competing for the same funding every year.

But what are all of these independent organizations doing? Are there really 1,212 distinct and unrelated social problems to be solved?

Probably not.

Following the 2010 earthquake in Haiti, the amount of duplication between nonprofit silos became painfully obvious. Nine hundred separate volunteer and government organizations simultaneously descended on the island with workers and aid. Although everyone had the right intentions, when so many people from so many discrete organizations showed up, there was no way to get them all into the severely damaged ports and airstrips. In many cases, untrained and unskilled volunteers were allowed into the country ahead of badly needed doctors, nurses, and rescue workers, who were redirected by the military to nearby islands to wait until congestion cleared. To add to the chaos, new organizations were sprouting up by the minute on the Internet, collecting funds and recruiting even more volunteers to help. Rather than optimize organizations such as the American Red Cross, which has been successfully delivering emergency services to disaster victims for 130 years, each independent silo charged ahead on its own, adding more complexity, redundancy, confusion, and delay to an urgent situation. The lack of collaboration with experienced organizations like the Red Cross resulted in nonprofits tripping over one another to administer aid. They fell into the trap of believing that independent efforts were somehow superior to collaboration.

But the facts suggest otherwise. A preponderance of silos is like having too many cooks arguing in the kitchen, despite all having recipes that taste exactly the same. A person could starve waiting to eat.

And what is true for health care and nonprofits is also true for business.

Dr. Carol Kinsey Goman, business consultant and author of *"This Isn't the Company I Joined": How to Lead in a Business Turned Upside*

Down, reports that "a study by *Industry Week* found that business functions operating as silos are the biggest hindrance to corporate growth. A more recent American Management Association survey shows that 83 percent of executives said that silos existed in their companies and that 97 percent think they have a negative effect."

As complexity mounts and facts and knowledge become more difficult to obtain, silos make acquiring vital information that much harder. If we have to scale the Great Wall of China to get to the facts, is it any wonder we're willing to settle for counterfeit correlations, best guesses, and *beliefs?* Beliefs are not only cognitively easier to understand and accept as true, but there's also no need to rappel a steep slope and no one waiting at the top to push us back down again.

Moreover, scaling a silo wall to collaborate is often an exhausting, frustrating endeavor riddled with danger. Collaboration is difficult. And collaboration in a complex, fast-moving environment is often impossible.

How do we fix it?

We have to start by tearing down our own silos first. Like E. O. Wilson and James Watson, we have to assume that people who are striving for the same objective are better off sharing information and resources than they are competing with one another. We have to be willing to put primitive instincts to protect territory aside and join forces for the greater good of humanity.

The Trouble with Territories

Biologists who study chimpanzees—with whom we share a great deal of our genetic material—claim that *the occurrence of silos is nothing more than a natural extension of territoriality*. Once again, the explanation for why it has become difficult for groups to work together stems from instincts that were once vital to the survival of our ancestors.

Territoriality is the process by which animals establish boundaries in order to protect food, water, mates, and their young in order to assure survival. Though the size of protected areas varies from species to species, as does the degree to which territorial borders are clearly defined, chimpanzees and other animals, including humans, are known to quickly establish and defend territorial boundaries for their own welfare. This is the reason we built moats around castles. It's also the reason we have controversial immigration laws. It's the thinking behind erecting decorative fences to separate our property from our neighbor's and why we distance ourselves from coworkers who transition to another department or go to work for a competitor despite once calling these people our friends. They are all attempts to safeguard our survival from "outsiders."

From a historical perspective, as humans made the transition from nomadic hunters and gatherers—people who were not tied to defined territories—to relying on agriculture, the need to defend terrain that was favorable to growing became stronger. Protecting turf was tantamount to protecting the food that was needed to stay alive. Boundaries became more rigid and the penalties for intrusion grew severe. Crossing a boundary without permission meant imprisonment, and sometimes even death.

In a complex social environment silo thinking may be nothing more than an irrational instinct to defend "territory" to improve our opportunities for survival. After all, an attack from a coworker, encroachment from another department, a criticism from our boss, even an opposing ideology from a competitor, all represent threats to our job, our status, our ability to make a living and provide for our offspring. In this way, silos not only simplify complexity by breaking it down into smaller, understandable components, they define "social territories" which must be vigorously protected.

Psychologist Aidan Sammons summarizes how silos work in this way:

> In order to impose order and predictability on a complex and unpredictable world, we simplify it. Simplification is achieved via abstracted mental models. A social-cognitive perspective on territoriality would suggest that the division of the world into primary, secondary,

and tertiary territories is one such mental model, used by people to generate expectations about, understand, and predict the behavior of others (Edney, 1975). Because we have control over our primary territory, we can predict what is likely to happen there.

Today, however, our prehistoric instincts to defend turf no longer serve society. In fact, silo thinking and behavior pose more risk than rewards as complexity mandates cooperation within and among disparate groups.

Secret Solar Silos

Sometimes, as in the case of health care, disaster relief, and business, the effect silos have on impeding progress is obvious. However, there are many more examples in which powerful solutions to our biggest challenges never come to light.

For example, when it comes to free solar energy for every household on the planet, we may have already solved the problem.

Surprisingly, the solution didn't come from a venture-backed start up in Silicon Valley, a university laboratory, or the Department of Energy. Instead, it comes from the most unlikely of places: the National Aeronautics and Space Administration (NASA).

NASA was signed into existence in 1958 by President Dwight D. Eisenhower "as a result of the *Sputnik* crisis in confidence." It's initial goal was to achieve human space flight, which led to the Mercury and Gemini projects, and eventually culminated in the Apollo space program and putting the first man on the moon. NASA then went on to develop *Skylab* and the Space Shuttle—two highly successful examples of complex collaboration.

Over time, however, America's love affair with outer space began to cool and NASA began to worry that its work was becoming irrelevant. In an effort to prove its commercial viability, the space agency entered a period of commercial partnerships wherein much of its focus was on communication satellite technologies, GPS navigation, and Landsat and Earth Observing. But as the competition for funding in Washington, D.C., heated up, NASA grew increasingly insecure about its role in the twenty-first century. How many take-offs and landings of the space shuttle would Americans watch?

So, NASA turned their attention to a new market, one the country was growing increasingly concerned about: renewable energy. In an effort to restore NASA's once-prestigious leadership role in the world, a handful of scientists began a secret program: *space-based solar power*. They would solve the problem of unlimited clean energy for the entire world once and for all.

When we stop to think about it, the most efficient place to collect solar energy isn't on the surface of the planet. The atmosphere acts as a shield that protects us, but this same shield also greatly reduces the strength of solar energy that can be captured.

In outer space, however, there is no atmospheric interference, so the efficiencies are magnitudes greater than what we achieve by laying solar panels on our roofs.

For decades NASA has been experimenting with solar cells in outer space in order to power satellites and spacecraft, and during this time a small group at NASA has also been perfecting a method for capturing energy and delivering it safely back to Earth's surface.

Electricity from satellites in outer space?

Sounds like science fiction.

Imagine the impact if every household had something as small as a satellite dish (akin to satellite television) by which we could receive all the power we need—for free. That's right: Free solar power could be reformatted and beamed safely into every home. Not only would it help our pocketbooks, but it would also mean no more utility plants or giant towers with cables strung across the desert. No underground trenches, nuclear waste cemeteries, or huge carbon emissions from coal-fired plants. The nation would be more secure because then there would be no centralized utilities, no major power lines to target that could cut off power to critical functions.

Space-based solar would change everything.

So what's stopping us?

It's shocking to discover that the scientists who have been working on space-based solar energy at NASA have been banging on the door of the U.S. Department of Energy for over a decade.

But no one would answer.

NASA? Aren't those the guys who invented Tang?

Like the CIA, which tried to make data available to academia for environmental research, green energy was far outside of NASA's stated mission. NASA was accused by the DOE of "mission creep" and ordered to stick to space exploration.

No matter how many times the scientists at NASA tried, they were unable to break through the silo walls that separated energy from space research. Meanwhile, billions of dollars were being invested by the DOE and Cleantech venture capitalists in technology that NASA knew was inferior to what they had already proven would work in their laboratories.

But the scientists at NASA were government employees bound by strict confidentiality. What could they do?

Frustrated and defeated at every turn, a handful of researchers went to work on tearing down the silos that were preventing progress. It was a risky endeavor and one that jeopardized thirty-year careers at the space agency: The scientists asked for permission to open discussions with Canada on a "joint research project" that would allow them to begin testing their discovery and prove its viability.

The Canadian government was all ears. In their eyes it was a chance to perfect space-based solar, put it into commercial use, and *then sell the power back to the United States!*

Suddenly the saying "You can't be a prophet in your own land" takes on new significance. After pouring millions of taxpayer dollars into inventing space-based solar energy—the permanent solution to unlimited, clean, and safe energy for the entire planet—the United States stands on the verge of allowing other nations to eclipse it, all because one U.S. agency can't get another one to listen. All because humans are still hardwired to defend their territory even when it is to the detriment of the greater good. All because of silos.

When this story breaks, some people may direct their outrage at the NASA scientists. Others will accuse the Department of Energy of being inept. Still other extremists will aim their sights at President Obama. There may even be individuals who will accuse me of being unpatriotic for going public with the fact that we have the technology to deliver unlimited energy from outer space. But it would all be misplaced blame and therefore unhelpful. Fortified government silos that are unable to cooperate, share information, and solve complex systemic problems together are the problem—not individual players.

When you think about it, space-based solar may be an alarming example of how silos prevent progress, but is this example substantially different from continuing to readmit almost 40 percent of the ER patients within 90 days? Or the resistance to sharing information among the CIA, NSA, FBI, and Homeland Security? The historical battle between genetics and evolution in biology?

Convergence Instead of Competition

In his 1998 book *Consilience: The Unity of Knowledge,* E. O. Wilson explained that silos have a more insidious effect than preventing a few problems from being solved here and there. Wilson warned that "professional atomization" also works against unifying the cumulative knowledge, discoveries, and science we now have at our disposal. Whether it's black holes in outer space or the current global recession, Wilson argues that thinking in silos prevents us from leveraging *all* the known laws in physics, music, chemistry, engineering, economics, and biology together to explain natural phenomena. In his view, the barricades that stand in the way of centuries of knowledge must be torn down in order for humanity to progress.

Wilson has it right.

The more fortified and numerous silos become, the further away humankind strays from a unified, systemic approach to our greatest threats. Think of humankind as garrisons of cognitively limited soldiers marching against quickly advancing complexity. Now imagine what chance these garrisons have when they don't work together to

outmaneuver a formidable aggressor. Success requires a level of cooperation and unification that we have yet been unable to achieve.

So what will it take to shift from silo thinking to collaboration? To
pave the way for systemic solutions? For the DOE to bring NASA into
the fold?

Here writer Saul Kaplan gets the final word: "It is not the technology that gets in the way of innovation. It is humans and the organizations we live in that are both stubbornly resistant to experimentation
and change. If we want to make progress on the big issues of our time,
we have to look up from our silos and become more comfortable re-
combining capabilities in new ways in order to connect with the unusual suspects."

— 8 —

Extreme Economics

The Fifth Supermeme

Every person I know has a strange relationship with money. They want more of it. They spend too much of it. They invest, inherit, protect, and live in fear they'll run out of it. Some people never talk about money. And some can't quit talking about it. Marriages break up over it, children are spoiled by it, and celebrities empowered by it.

But mostly we wear money on our shirtsleeves where we once wore our hearts.

Not long ago, there was a time when we put our faith in each other instead of legal contracts, the stock market, and social security—a time when the single barometer of success and failure wasn't money.

Other values still mattered.

But today there is a growing feeling that money has stopped working the way we intended it to, an inkling that our romance with economics may have gone too far.

After all, money isn't really the reward for hard work.

If it were, gardeners who toil day after day in the hot sun would be wealthier than executives. And how about farmers in Africa and China who work within an inch of their lives to grow just enough to eat? Do they work less than Bill Gates?

Money isn't a consequence of being lucky either. If it were, people who duck hurricanes and survive car accidents would be rich.

And money isn't the result of being smarter. A lot of academics I know barely scrape by.

People with money aren't harder working, luckier, or smarter. Likewise, people who barely make ends meet aren't lazy, stupid, or unlucky sad sacks. But in a commerce-driven world, this commonly held stereotype—the belief that people with money are in some way superior—is just as harmful to human progress as racism, sexism, and ageism.

So, it must be said: Money is just money—paper and round pieces of metal invented by humans to make transactions easier. Even though wealth doesn't have any intrinsic characteristics except those we attach to it, we are awfully quick to associate being rich with a battery of desirable traits. As a result, acquiring wealth has become the new shortcut for obtaining these traits. Instead of being a natural outcome of accomplishment, for the first time in history money has become the end goal.

A Two-Sided Coin

The fifth and final supermeme is called *extreme economics*—and there's a reason we saved it for last.

The economics supermeme occurs when *simple principles in business, such as risk/reward and profit/loss, become the litmus test for determining the value of people and priorities, initiatives and institutions.* We begin uniformly applying the strategies we use to succeed in business to the other areas of life. In other words, in a business-focused society, financial success separates winners from losers and sure bets from pipedreams, irrespective of whether we are buying a family home, choosing a mate, or saving the planet.

Unlike the first four supermemes, *irrational opposition, the personalization of blame, counterfeit correlation,* and *silo thinking,* which are easy to see as harmful, *extreme economics* feels more like a relative who came to visit and stayed too long. We have mixed feelings.

On the one hand, many people argue commerce is good. The pursuit of profit has been the engine behind vast advances in modern technology and science. They point to an improved standard of living, longer mortality, and new heights in efficiency and productivity—evidence that profit provides a powerful incentive and is therefore an essential component of progress.

And for the most part they are right. With the fall of the Soviet Union and other communist/socialist nations, progress favors societies committed to the free, unbridled pursuit of wealth.

On the other hand, it can also be argued that too much focus on the financial bottom line prevents many solutions that would benefit humankind from coming to fruition. When economic considerations become *the only* considerations, we're on a slippery slope. Helpful remedies that cannot be financially justified never get traction. Just ask Dean Kamen.

Remember Kamen? The iconoclast and inventor of the Segway scooter? Typical of many geniuses, Kamen sees opportunities for improvement everywhere. Recently he shifted gears from futuristic wheels to global ideals: *Kamen now believes he has solved the world's potable water problem once and for all.*

Not a mitigation, mind you—a real and permanent cure.

To put Kamen's breakthrough in perspective, a recent article in *Newsweek* magazine revealed that "water-related diseases kill a child every eight seconds and are responsible for 80 percent of easily preventable illnesses and deaths in the developing world." Furthermore, additional statistics gathered by the United Nations show that well over one billion people around the world are currently without clean water. And this figure is rising as industrialized nations such as India and China join the ranks of countries that have endured contaminated water for generations and for which the problem is worsening.

In 2008, Kamen's company announced a small water purifying system, called the Slingshot, that "produces 10 gallons of clean water an hour on 500 watts of electricity." That's roughly the amount of electricity needed to power "one string of Christmas lights for about an hour," so according to Kamen the system can be easily fueled by methane gas produced from a small amount of cow dung. Unlike other water purifiers, Kamen's invention requires no consumables such as chemicals or activated charcoal or osmosis membranes. It also cleverly captures some of the heat used to purify the water and uses that heat to offset the electricity needed to operate the purifier—a feature inspired by the 1816 Stirling Engine.

Kamen estimates that he is within two years of a high-volume unit that can produce enough potable water to "meet the needs of the entire world."

Kudos for Kamen.

Despite his breakthrough, Kamen is the first to admit that solving the problem in the lab is only half the battle. The trick is getting it in use, which requires convincing the public that the long-range benefits outweigh the short-term price tag. And it's here that Kamen stumbles: "We believe each machine will be under a couple thousand bucks. We need to develop the business models and the relationships. In some countries it's going to require microfinancing and entrepreneurs; in other countries it'll be nongovernmental organizations or governments."

Good luck.

For decades, we have had products that can provide every individual on the planet unlimited free energy. The same goes for food and vaccines and medicines. What's more, we could have an immediate effect on global warming if we were willing to paint the roads and roofs white, if we stopped chopping down more trees than we plant, or if we were willing to give up burning coal. We already have the knowledge and solutions to lessen many of our stubborn problems—and even in some instances to cure them. The problem isn't whether we can come up with incrementally helpful solutions. The problem is what gets in the way of deploying them—and when it comes to solving a systemic problem, there are dozens of roadblocks. Among these roadblocks, profit—not technology—is one of the biggest.

That's because broad systemic solutions that benefit humankind don't always fit accepted economic models. And when they don't, progress is inhibited.

Unless Kamen and others can turn their innovations into a financial boon, the Slingshot and innovations like it aren't likely to be adopted on a scale large enough to have any impact. Just ask solar panel manufacturers who have been peddling their wares for four decades now; or hydroponic farmers who produce gigantic crop yields using only a fraction of the water needed by conventional methods— all without soil; or nutritionists who can prevent an encyclopedia of human illnesses by simply altering a person's daily diet. Having consulted with hundreds of start-up companies over the course of my career, I have witnessed one helpful technology after another fall by the wayside because the financial upside simply wasn't large enough to attract funding. Sadly, more important breakthroughs die on the vine because they cannot be economically justified than ever see the light of day.

There's no doubt about it, in the twenty-first century, profitability has become the most powerful barometer of legitimacy.

But think about it. What does profit really have to do with fixing a global drought or curtailing the spread of fast-moving viruses that cross national borders in a matter of hours? How do we quantify the return on investment of upgrading school curricula or making prisons habitable? Do we have to prove the solutions to these problems are a sound investment or that money is being put to good use? Is it even possible to do so?

And therein lies the irony. The same economic incentives that spur entrepreneurial innovation also quell some of our most vital and necessary discoveries.

As the problems a civilization must solve become more complex, more global, and more systemic, to some extent the bottom line becomes irrelevant. Though the principles that govern economics are rational and designed to assure a positive outcome, when these same principles are applied to complex global problems, they become counterproductive. Using risk/reward to measure the value of a global, humanitarian solution is a lot like using a ruler to measure intelligence. Wrong tool.

Ecumenical Economics

We may not always use the same language as an economist, but most of us think like one. Today, when faced with an important decision—whether it involves illegal immigration, national health care, or getting a divorce—economists, politicians, executives, and the man on the street all run through a similar checklist:

- *Capitalization:* How much is this going to cost me?
- *Risk:* What's the probability the venture will fail, and how much will that cost?
- *Return on Investment:* Is this the best I can get for my money?
- *Leverage:* Will this investment lead to achieving other goals?

To prove the point, we need only look at how popular prenuptial agreements have become within the last two decades.

The reason prenuptials have recently exploded is that we now view marriage not only as a romantic union but also as an economic partnership. Similar to the ancient practice of dowries or marriages that were once motivated by political alliances, prenups are modern contracts designed to define the financial terms of marriage in advance.

They have become so prevalent in the past few years that most of us can't imagine someone like Donald Trump, Bill Gates, or Madonna getting married without one. In fact, the more assets a person has, the more they should worry about protecting those assets. But considering this from a nonbusiness perspective, why in the world would someone marry a person they can trust with their life, their children, their future, and their heart, but not their money?

Even more to the point, why don't we ever hear about anyone signing a precustody agreement? Presumably, most couples who marry either plan to have children or bring their existing offspring into a new family. With this in mind, why don't we have contracts that spell out how the children will be cared for in the event of separation? As long as we are taking advanced precautions to lower risk, wouldn't this make sense? It would if we didn't put the welfare of our assets ahead of the welfare of our children.

Then, when a marriage is over, what happens? Each side attempts to get the best *return on their investment!* Ask anyone who has gone through a divorce. Most divorces become contentious, elongated negotiations over a property settlement, alimony, and support—a process that can often last years and involve multiple lawyers and specialists, depending on the size of the estate. Here again, the debates over child custody are frequently reduced to a debate over child "support."

But prenuptials and divorce settlements are just two modest examples of the encroachment of economics on our lives. Everywhere we turn, someone has figured out how to make or save a buck. What used to be plentiful vistas of undisturbed lands are now planned "open space" or national parks with toll booths. Need a blanket on an airplane, pay a dollar. Park your car? It will cost you. Want to dig a well in your backyard? Write a check for a permit and stand in line.

But it's important to note that this mentality is a new phenomenon that didn't exist thirty years ago when campgrounds were free, we didn't view our homes as a financial asset, and no one ever heard of a prenup.

The Struggle for the Greater Good

If the pursuit of profit is the engine behind innovation, then should we also assume it's the engine behind serving the greater good?

Not exactly.

Here's one small but telling example that demonstrates how one doesn't always serve the other.

In a disturbing 2010 report, a young woman discovered that New York Wal-Mart and H&M stores were throwing large bags of brand new, unsold goods into trash dumpsters behind their stores—but not before taking scissors and deliberately cutting holes in them to make them unusable.

According to Cynthia Magnus, who found the ruined clothing, "[There were] gloves with fingers cut off. . . . Warm socks. Cute patent leather Mary Jane school shoes, maybe for fourth graders, with the instep cut up with a scissor. Men's jackets, slashed across the body and the arms. The puffy fiber fill was coming out in big white cotton balls."

But why?

Each year the percentage of people living below the poverty line in New York City hovers around 20 percent. During the cold winter months, the poor suffer harsh conditions on the street. So what would cause a profitable chain to cut up warm coats and gloves rather than offer them to nearby charities?

The extreme economics supermeme.

After all, once word got out that customers could get the clothing for free, what would this do to prices? Future sales? Customers?

From a *factual* standpoint, probably nothing. The homeless weren't shopping in these stores, so they weren't going to lose them as customers. Likewise regular customers weren't likely to line up outside a charity, hoping on some slim chance they might be able to get the exact same merchandise, in their size, for no money.

But as we already established, that's not the point. It's what they *believed* would happen to their bottom line that mattered: better to make unsold goods "unavailable" than compete with yourself. And there we have it: At the heart of every irrational behavior lies a disturbing supermeme. In this case, rational business principles stood in the way of doing what was in the interest of the greater good.

Today, economic considerations overwhelm other values to such an extent they now have the power to decide how and if humankind's greatest threats will be solved.

In 2009, on one of my visits to Harvard University, I ran into E. O. Wilson just after he returned from one of the prestigious annual conferences at which he is regularly invited to speak. When he accepted the invitation, he assumed they wanted him to discuss a topic on which he is a known expert. Wilson is the most famous naturalist in the world. He is also the acknowledged father of sociobiology and biodiversity, the undisputed authority on ants, a prolific writer, and a passionate advocate for the green movement. But Wilson returned dismayed and deeply concerned by what had transpired:

> All anyone wanted to talk about was the recession. What was going to happen? What caused it and how long was it going to take to recover? Supposedly the smartest people in every field were there. The best and the brightest humankind has to offer. And we accomplished nothing. Not one thing. In fact, they wanted to know if I would mind changing my talk. They asked if I could talk about the *recession* since that was what everyone was interested in.

Then he paused before adding, *"Economics is ruining everything."*

It's a sad commentary when critical resources such as E. O. Wilson and other experts lose their ability to share information and collaborate on critical problems—especially at a time when we need all the great minds we can muster to overcome the dangers that lie ahead.

Sadly, Wilson's experience wasn't unique. I have heard the same concerns echoed by other authorities who have been attending global summits for decades. They worry that nothing of any substance is getting done because economics has hijacked every other agenda. And with this comes the growing concern that civilization is driving toward some dangerous singularity.

Why is everything we do suddenly governed by economics? How did these principles get a stronghold on civilization? Why have we become obsessed with money? Once again, the answer can be traced to our early ancestors and inherited biological predispositions.

A Chilling Change in Chimps

A few years ago, graduate students at a major university devised an experiment to see what would happen if chimpanzees were introduced to a few basic principles in economics.

The idea behind the experiment was to teach the chimpanzees how to become consumers and observe their behaviors as they completed different types of transactions. To accomplish this, every week a group of chimpanzees were each handed a set number of tokens. The tokens represented money that could be used to make purchases just like regular consumers.

As soon as the tokens were distributed, one chimpanzee at a time was allowed into a separate cage adjacent to the compound where the chimpanzees lived as a group. A lab assistant would enter the private cage with a tray of snacks that the chimpanzee was allowed to examine and purchase using his tokens. After a chimpanzee made his selection, the lab assistant would help him count out the correct number of tokens, place the tokens on the tray, and hand over the snack.

When the chimp was done with his shopping, he rejoined his group in the community cage. Then the next animal was allowed into the private area to make his purchases. This was repeated again and again until all the members of the troupe had an opportunity to shop.

But there was one small snag.

All the snacks on the tray had different prices. So, whereas one token would allow a chimp to purchase five crackers, that same token was worth only one small piece of an apple. Each chimpanzee had to determine for himself which snacks were worth spending his tokens on.

Within a remarkably short period of time the chimpanzees learned the value of their tokens through basic trial and error. They soon became discerning consumers who made humanlike judgments about value and price, often carefully examining the items on the tray before indicating their final decision to the lab assistant.

As the chimps grew more familiar with the snacks and pricing, transactions were completed more quickly. In advance, the chimpanzees had already determined the best "value." So, when they entered the private cage, they quickly plopped down their cash and exited with their purchases without so much as inspecting the now familiar goods.

But then the researchers decided to mix it up a little.

They changed the prices of the snacks.

Overnight, they "inflated" the number of tokens needed to purchase the chimp's favorite treats so the tokens were now worth much less. When the chimpanzees discovered they weren't getting as much for their money, there was noticeable confusion, followed by agitation and rage. They rebelled against the lab assistants, frequently becoming belligerent and uncooperative. Some went on strike and stopped buying snacks altogether. Others grew indecisive, aggressive, and depressed.

Conversely, when the scientists lowered the prices, the chimpanzees showed great delight in their sudden windfall. The surplus of goods was reflected in their generosity toward other chimps, fewer squabbles, and, sometimes, lethargy.

But what ultimately caused the university to close down the study were the serious side effects no one could have foreseen. As the chimpanzees began applying their newly acquired principles of economics to other areas of communal life, unexpected and disturbing behaviors emerged.

For example, in a surprising turn of events, the researchers observed the female chimpanzees flirting and offering male chimps sex in exchange for tokens.

Not snacks mind you. Tokens. Real cash.

It was a shameless display of prostitution, and it appeared to be motivated by only one thing: the female's desire to acquire wealth—and they were willing to do whatever they needed to, to get it.

As if that wasn't amusing enough, one day there was a heist.

A normally shy, well-behaved chimp entered the private cage preparing to purchase his snacks. He pretended to look over the selection, strategically positioning himself as close to the food and tokens as possible. Then suddenly, in an abrupt gesture, he slammed the bottom of the tray, and the food and tokens flew into the air and landed in the communal cage. The nearby accomplices screamed and quickly grabbed the loot before the lab assistants could get control of the pandemonium.

No doubt about it: It was a premeditated robbery.

In addition to prostitution and robbery, a battery of other unusual behaviors began appearing: bartering, manipulation, hoarding, even the primitive beginnings of a black market for pricey snacks.

Then animal welfare advocates got word of the experiment and became outraged. In their view, the chimpanzees were being corrupted and the bizarre behavior among the animals was all the proof they needed. To avoid further controversy and the prospects of bad publicity, the university was forced to quickly shut down the experiment. Peace was restored to the small chimpanzee troupe that, for a brief moment in time, had become expert consumers. Overnight, there were no more tokens, no more snack store, and no more crime or sex for money.

But not before we had an opportunity to get a close look at *extreme economics* in action.

Monkeys and Money

Author and renowned biologist Richard Dawkins provides an evolutionary explanation for the impact money briefly had on the small chimpanzee troupe as well as on humankind today. In his landmark book *The Selfish Gene*, Dawkins writes: "Natural Selection favors individuals who successfully manipulate the behavior of other individuals, whether or not this is to the advantage of the manipulated individuals."

Nowhere is successful manipulation easier to spot than in economics.

From a biological standpoint, the most successful accumulators of wealth greatly increase their survival opportunities by building surpluses. So, when opportunities to increase wealth present themselves, our normal biological response is to capitalize on these opportunities to improve our survival. It's a hardwired instinct.

From this perspective, monkeys offering sex for tokens is perfectly natural, as are robbery, hoarding, cheating, and aggression. The tenets of Natural Selection dictate that each organism must do its best to manipulate the environment to its advantage, and when a dominant feature of the environment is money (tokens), we instinctively manipulate economics to our favor. It's that simple.

Harvard researcher Terence Charles Burnham states it another way: "Evolutionary theory predicts that, subject to physiological and informational constraints, organisms will act to *maximize* their reproductive success." Put in this context, it is easy to understand why the principles of business have become so pervasive. *Economics is the perfect, most efficient system for manipulating resources to our evolutionary advantage.* Money gives us access to health care, clean water, food, safety, shelter, and other advantages vital to modern survival.

It's no wonder we try to leverage economic principles in areas of life outside of business. Once we learn to successfully manipulate commerce to our benefit, it's only natural to endeavor to use these same tools to our advantage elsewhere.

In the case of the chimpanzee troupe, money (tokens) was introduced by curious university researchers. But in the case of humans, how did money originate? How did it come to infect human behavior?

Man, Money, and Mayhem

Today it is difficult to imagine how societies ever functioned before the invention of currency or credit. But to truly understand the origins of the economics supermeme—how money and commerce came to dominate modern behavior—it's useful to take a brief journey back in time, beginning with the simple origins of money.

Prior to the existence of currency, the process ancient civilizations used to exchange goods and services was called "barter." Barter was an effective way to trade because it was based on the value of goods and services with which we were already familiar. In those times, exchanges involved manual labor and products like produce and livestock.

In a barter economy, if you had more cows than you needed, and I had more wheat than I could consume, we would simply agree to an amount we *both* felt represented a fair exchange. Then we would make a trade—one of your calves for some of my wheat. That was that. No refunds, no membership reward points, no cash-back coupons.

Barter, however, depended on what economists call a "coincidence of wants," where the cow and wheat had to be available for exchange at the exact same moment in time. So if you needed my wheat today and your cow hadn't given birth to the calf you wanted to trade, you were out of luck.

The obvious solution was to trade the wheat for some "intermediate commodity," which would allow you to receive the wheat and then later sell me the calf once it was born. This and the convenience of carrying paper and coins to market instead of wagons full of wheat and cattle were the original impetus behind the creation of money.

However, determining the values of livestock and produce for trade is considerably cognitively easier than determining the value of currency. Currency is based on a sophisticated process of associating some value to a paper or metal "symbol" of that value. In other words, the paper and coins themselves aren't worth much: It's what they *represent* that's important. And assigning a value to an object that represents something else is more much more complex than assigning a value to a real product such as a calf or a bushel of wheat. In this way the widespread acceptance of a "symbol" of value—money—represented a major leap in human advancement.

Once humans began mass-producing currency, what we know as modern economics was not far behind. Intricate laws and systems for credit, debt, investment, speculation, and principles of commerce quickly evolved. From there, it didn't take long for the principles of economics to become inseparable from governance. Bear in mind that in earlier times, when people bartered directly for goods and services, value was established by individuals, not governments or experts. However, once the production of currency became the centralized providence of governments, humankind abandoned barter and entered a new era in economics. Values became much more obtuse—and so did the meaning of wealth. Wealth became symbolic.

Then came the industrial revolution.

During the industrial revolution, unprecedented advances in mass production were achieved that resulted in vast surpluses of cheaper products. In a short period of time we transitioned from a primarily agrarian society to an industrialized one in which supply could be ramped up to fulfill virtually any level of demand.

The next step was to *optimize*—to produce goods at an even lower price, sell more, expand markets, squeeze competition, merge, subcontract, diversify, form partnerships, leverage, brand, position, and promote. This marked the beginning of *extreme economics*: the advent of sophisticated principles aimed at maximizing profitability.

Extreme economics has resulted in achieving new heights in efficiency, institutionalized entrepreneurship, and sophisticated methodologies for enhancing profitability by manipulating resources, people, and markets. This new epoch has been responsible for the formation of venture capital organizations that single-handedly financed a revolution in electronics, computing, the Internet, and cellular communications. Extreme economics also led to an explosion of business schools and MBAs as well as the birth of a new breed of new financial icons like Peter Drucker, Jack Welch, Warren Buffett, and Donald Trump. Extreme economics has put pressure on governments, public schools, and nonprofits to balance their budgets, increase their output, improve asset management, and develop competitive strategies. What's more, extreme economics has also been the impetus behind standardizing international trade. The drive toward greater efficiency has caused all industrialized nations to adopt remarkably similar business principles.

In these ways, the era of extreme economics perfected the advances made during the industrial revolution by elevating economics to a universally accepted supermeme.

But with all of this progress, there comes a downside. Our monetary systems have become so complex that the most brilliant economists in the world, such as Alan Greenspan, now have difficulty describing exactly how the value of money is determined anymore. Today, the value of currency depends on many things: production, exports, circulation, liquidity, foreign currencies, inflation, deflation, and a hundred other factors. In short, it's extremely complex.

When the complexity of our monetary systems exceeds our ability to acquire or understand the facts, it is normal for us to begin relying on *beliefs* instead of proven knowledge.

And when it comes to modern economics, there is no belief more dangerous than our attitudes about credit.

Nowhere has the influence of money on human behavior been so similar to the chimpanzees than when it comes to the obscure notion of credit. From a biological perspective, credit provided another powerful tool for manipulating our environment: It was a way to acquire more goods, faster, while appearing to have no negative consequence. As a result, our natural biological instinct was to leverage credit to our favor, and, as we have recently learned, many people, businesses, nonprofits, governments, schools, and organizations did just that.

The more credit that was offered, the more we wanted. Drive a new car off the lot with nothing down? Yes! Buy a new house with no assets or income? Yes! Get 30 percent off your purchase if you apply for a store credit card? Yes!

So it comes as no surprise there is a great deal of confusion today about what we own and what we owe. Many young people now treat credit cards the same way they do cash. Instead of viewing credit as a loan that must be repaid, they see it as indistinguishable from their net worth. In their minds, a higher credit limit is akin to depositing more money into their checking account. And with a new generation of debit cards tied to their checking accounts, is it any wonder that young people are more confused than ever?

But the proof is in the pudding. In 2007, for every $135 the average American saved, they acquired approximately $9,800 in personal debt. In other words, every man, woman, and child in America now owes more than seventy-two times more than they put away for a rainy day.

And if that sounds ominous, we run our governments the same way we run our households.

In 2009, the federal debt in the United States was reported to have passed the $12 trillion mark. But this doesn't begin to tell the full story. Unfunded obligations, commonly called "entitlement programs," such as Social Security, veteran's benefits, unemployment compensation, Medicare, Medicaid, food stamps, and agricultural subsidies, are mandated by law and, therefore, also must be paid. According to the Congressional Budget Office, "The present value of these deficits is approximately $41 trillion."

Now add in the deficits accumulated by individual states and counties, estimated at $30 to $50 billion.

The grand total of the federal deficit, mandatory entitlement programs, and state and county obligations? Over $53 trillion, or about $175,000 for every man, woman, and child living in America.

Tack on $9,800 of personal debt, and each of us is on the hook for just under $200,000 apiece—with interest piling up every day, a family of four will soon be pushing $1 million of personal and government debt that must someday be repaid.

Yet despite these alarming figures, we keep borrowing. We keep leveraging our future as if the bill will never come in the mail.

Then suddenly, it caught up to us in 2007, when a record number of people could no longer pay their mortgages. A large number of homeowners succumbed to the weight of debt that had been accumulating for years. What's more, experts and the heads of the largest financial institutions in the world seemed caught by surprise. *They, too, had succumbed to the false belief that credit was the same as bankable assets.* It was as if suddenly everyone, all at once, realized their assets were built on a house of cards. When the wind shifted, it all came tumbling down.

This is what happens when supermemes such as extreme economics overtake rational thinking and facts. We become susceptible to behaviors that, in the long run, cause great harm. Just ask the number of working families who forfeited their homes because they lost track of the difference between cash and credit. Was their desire to acquire a home they could not afford so different from the chimpanzees who were willing to steal and trade sex for more snacks? Like the chimpanzees, we are an organism that is biologically predisposed to acquire as much surplus as possible to increase our survivability. And it turns out, we aren't as particular as we think we are about how we acquire those surpluses.

Perhaps the animal welfare advocates were right to shut the university's experiment down before the intrusion of commerce led to more dangerous consequences. We'll never know.

Pendular Public Policy

In addition to affecting our personal lives and once-immune social institutions such as universities and matrimony, the extreme economics supermeme has an equally powerful impact on public policy. As the financial bottom line increasingly becomes the defining measure of success, good business can easily become confused with the greater good of civilization.

This confusion is especially clear during economic downturns.

For instance, in 2009, when faced with a historic deficit, the State of California found themselves forced to reconsider unconventional sources of funding they had vehemently rejected in the past.

As the world fell into a deep recession, California, the twelfth largest economy on the planet, was forced to start issuing IOUs instead of paychecks to taxpayer, lenders, and local governments. State tax refunds were also delayed, and many government services were closed or cut back to skeletal levels. Fire departments were consolidated, police were laid off, and public libraries were closed. As the recession grew worse and the number of unemployed, bankruptcies, and foreclosures climbed, it became clear that increasing taxes on a citizenry already struggling to make ends meet was no solution for filling federal, state, county, and city coffers.

How would the state continue to provide support for essential services like fire and police protection? What could they do?

Then someone suggested legalizing marijuana.

If marijuana could be sold legally, then it could also be taxed, and this would generate an immediate revenue infusion. According to Alison Stateman, a reporter for *Time* magazine: "Pot, is, after all, California's biggest cash crop, responsible for $14 billion a year in sales, dwarfing the state's second largest agricultural commodity—milk and cream—which brings in $7.3 billion a year, according to the most recent USDA statistics. The state's tax collectors estimate the bill would bring in about $1.3 billion a year in much needed revenue."

In this same article, California State Assemblyman Tom Ammiano elaborated on the case for legalizing marijuana: "With any revenue ideas, people say you have to think outside the box, you have to be creative, and I feel that the issue of the decriminalization, regulation and taxation of marijuana fits that bill. It's not new, the idea has been around, and the political will may in fact be there to make something happen."

But why now?

Because when we add the huge influx of taxes to the $1 billion savings Orange County Superior Court Judge James Gray estimates would be saved each year from eliminating the arrest, prosecution, and incarceration of nonviolent offenders it's too tempting to ignore. Perhaps this is the reason that during the same week Ammiano introduced his bill, the U.S. Attorney General also announced that federal raids on marijuana growers, dealers, and users would cease, and each state would now make its own laws governing marijuana use. The federal government paved the way for states to legalize an illegal cash crop and did so right under the public's nose.

Keep in mind that being in favor of legalizing marijuana just one year earlier would have been the equivalent of political suicide. However, because the entire nation was in an economic crisis, politicians, to their credit, set aside their previous biases and stepped forward to embrace a radical new solution to their financial woes.

But then, just as quickly, the political climate shifted in the opposite direction as new facts came forward.

An economist pointed out that the original tax revenues forecasted for marijuana sales were highly inflated and would not produce the giant infusion originally anticipated. Early forecasts were erroneously based on the high price that *illegal* marijuana commanded. Once legalized, it was likely that marijuana would flood the market and thereby drive prices down. As soon as supplies increased and prices dropped, taxes from marijuana would look a lot like taxes from popular crops like strawberries and lettuce. In 2010, an independent study by the RAND Drug Policy Research Center indicated the prices would drop an astonishing 90 percent once marijuana was legalized.

Once the financial windfall was disproved, enthusiasm on Capitol Hill dwindled; however, a few advocates like Ammianno continue to argue there could be "substantial" economic advantages. The moral of the story?

Today, public policy and, arguably, society's ethics are easily shaped by economic need.

In many ways our attitude toward marijuana is no different than our attitudes toward afterschool music and sports programs, green energy, prison rehabilitation, and illegal immigration. If a solution to a problem makes good economic sense, we are in favor of it, but as soon as it is shown to be inefficient, costly, or devoid of any tangible economic benefit, we are quick to dismiss it. Think of it in this way: If we can't justify it economically, then it probably isn't such a good idea. Conversely, if it generates more revenue, it's got merit.

But is this really true?

Impairing Institutions

The effect of extreme economics doesn't only affect public policy; it also infects important social institutions. It doesn't matter whether the institution is marriage, a family business, or higher education—the financial bottom line now drives everything.

For instance, it's safe to say that in the minds of most people, universities and colleges still remain one of the last safe havens for scholars, freethinkers, inventors, and iconoclasts. Many of society's most helpful discoveries have emerged from brilliant minds that have sought the protection and resources of academia.

In fact, to protect the untarnished pursuit of knowledge, most advanced learning institutions adopted policies such as tenure, a process by which an expert is offered a permanent position on their faculty and enjoys protection from being terminated for his/her opinions. Most learning institutions recognize the important role that a neutral, fearless environment plays in the pursuit of the higher good, so they've taken great precautions to assure independence from outside influences.

Within the last three decades, however, this view has dramatically changed. Today, there is more attention paid to raising capital, expanding enrollment, and building a franchise than ever before, and important safeguards such as tenure are now under fire.

What's more, although a great deal of university funding still comes from the government and private donors, most colleges have now been forced to operate like for-profit businesses. Universities today invest more in building elaborate sports stadiums than they do in salaries because athletics is a big moneymaker. In addition, they are also quickly capitalizing on the Internet by offering convenient "online" courses and degrees. They stage elaborate concerts, festivals, and speaker's tours and carefully leverage hundreds of profitable campus concessions that have become as sophisticated as any national retail chain.

But by far the biggest boon in funding has come from partnering with big business.

Wayne C. Johnson, then executive director of Hewlett-Packard University Relations Worldwide, commented in 2004 that there was "an intensified need for collaboration between universities and corporations"—and he was correct.

Universities need more funding.

But why has collaboration with business suddenly "intensified"?

The answer lies in extreme economics: Collaboration means more leverage and profits for both institutions.

Today, it has become commonplace for the world's learning institutions to trade primary research for dollars—frequently becoming an extension of a public company's commercial research and development department. This partnership is successful because it provides funding to the university while allowing corporations to subcontract difficult, long-term, esoteric research to higher learning institutions. On the surface it's a marriage made in heaven.

However, as universities become more and more desperate, there is a side effect to the seemingly perfect alliance: Research that is commercially viable is favored over research that is not. And by favored, what I really mean is *funded*.

Look at what happened to Stanford University's and MIT's physics and engineering departments as the consumer markets for semiconductors, computers, and communications began taking off. Or the current partnership between biotech companies and the microbiology departments at Harvard, Cornell, and the University of Texas. In each case, dollars came pouring into these departments as soon as their research was shown to be commercially viable.

Sadly, if we want to preview which direction universities are headed, all we need to do is follow the stock market.

In a white paper titled "Intellectual Property: Universities, Corporations and Finding Common Ground," the American Society for Engineering Education (ASEE) summarized the dangerous trend in academic research as follows:

> Back in the late 1950s and early '60s, federal research spending for the physical sciences and engineering amounted to 2 percent of GNP. But that kind of spending ended in the 1970s, while research money for the health and biological sciences mushroomed. NSF [National Science Foundation] figures show that federal spending for engineering research remained fairly flat between 1970 and 2000 at around $5 billion to $7 billion; research money for the physical sciences hovered between $3 billion and $5 billion. During the same period, however, research money for the life sciences soared from about $5 billion to around $20 billion.

Today the idea of research for research's sake is slowly going by the wayside, and extreme economics is the driving force behind the unsettling change. Just ask the anthropology, zoology, or humanities departments of any university. These departments suffer regular cutbacks as they watch millions of dollars of business funding pour into engineering, biotech, and physics research. In short, every university's research budget has become inextricably tied to commercial ventures. As the ASEE article noted, "As many companies now use in-house labs primarily for product development and exploring advanced technologies; they are increasingly turning to research universities 'as a source of . . . applied research.'"

But it wasn't always this way.

Sure, there has always been *some* collaboration between education, the government, and business, but never to this extent: never to the point of business determining which departments and what research received funding; never to the point where the "return on investment" was the determining factor; and never to the point where discoveries that had no other value or purpose other than to benefit humankind were dismissed in favor of profitable endeavors.

But it is wrong to single out universities for bowing to the pressures of extreme economics.

The Directors of many nonprofits are also deeply affected by intrusive business principles. It is disheartening to see firsthand how much time the Boards and executives of nonprofit organizations now spend fundraising or figuring out how to deliver the same services for less money. Endowments have to be invested and managed properly, donors have to be courted, and nonprofits have to become marketing experts in order to successfully compete for donations. It's dog eat dog.

The focus on the bottom line also extends to church leaders, who are judged and rewarded by bigger, more prestigious assignments according to the size of the donations they bring in. Politicians, too, are rewarded according to the campaign dollars they help raise for their respective parties.

Everywhere we look, economics plays a dominant role in how institutions now operate. And the more we measure human progress in terms of dollars and cents rather than dollars and sense, the greater the myopia becomes. As remedies to the world's most complex and dangerous threats become limited to profitable investments, the full talents of the human organism never come to bear.

Fairness in Pharma

Perhaps nowhere is the conflict between economics and human progress more polarized than in the ongoing debate over health care.

When it comes to the trade-off between human life and profits, we are a society that clearly has two frames of mind.

To begin with, most health care and pharmaceutical companies are *for-profit* businesses. For the most part they are public companies. Many millions of people own health care and pharmaceutical stocks in their financial portfolios, which they depend on for income and retirement.

From this perspective, we the investors expect pharmaceutical companies to provide a healthy financial return. We want them to increase sales, lower expenses, and improve profits so the value of our stocks will go up. Otherwise, why buy their stock?

And so far, pharmaceutical, health care, and biotech companies haven't disappointed.

According to Pharmaceutical Research and Manufacturers of America (PhRMA), in 2002 pharmaceutical profits were five and a half times greater than the median for all industries represented in the Fortune 500. In September 2007 Med Ad News reported that the top twenty "Big Pharma" companies had a combined net income of approximately $110 billion dollars. Or put in more sobering terms, more than half of the government's loan to AIG.

In other words, in a little less than two years U.S. pharmaceutical industry profits alone could pay off the full AIG bail-out tab. This won't happen of course, but it does put those puny AIG bonuses we were once alarmed about into perspective.

But the fiduciary responsibility of pharmaceutical and health care companies to make profits for their shareholders also has an unsettling aspect to it—a sticky moral side.

How much should a for-profit company be allowed to charge for the one and only drug that may save a person's life? From a strictly economic perspective the answer is: *the highest price the greatest number of people are willing to pay.* It's a simple matter of supply and demand, and the demand for a life-saving drug is always high, so . . .

But somehow, when it comes to the question of life and death, we realize that's not really right. When children succumb to disease, even the most coldhearted among us demands that everything be done to relieve their suffering—at almost *any* cost. So what should a for-profit company do if it has to determine whether a person lives, dies, or suffers? On the one hand, the company is obligated by business ethics and the law to deliver the highest possible returns to their shareholders. On the other hand, the company has a moral responsibility to save lives.

Which is it? Profits or people? Our obligation to investors or responsibility to humanity? How does a company serve two masters?

The Pressure to Respond Quickly

Up to this point we have discussed the effects that currency, profit, and prehistoric instincts have on modern human behavior. But there is an equally powerful fourth by-product of the extreme economics supermeme: When business principles prevail, there is enormous pressure for individuals to respond to complex problems with great speed and efficiency. Quick, decisive action is prized over slower, thoughtful, methodical examination.

Although at one time five-year strategic plans were a necessity, today far fewer companies and countries take this exercise seriously, and many have abandoned it altogether. Being opportunistic, nimble, and able to react quickly to changing market conditions has become the name of the game in the twenty-first century. And when you consider how complex and fast moving the world has become in such a short period of time, can you blame us? It feels futile to develop long-term strategies when the global landscape is changing in nanoseconds. One minute General Electric is entering the booming computer industry, making rapid-fire acquisitions one after the other, and the next minute they are getting out. One day Microsoft and Google are on the verge of joining forces, and the next day they act as though they are competitors. One week a company proudly announces that it has exceeded analyst projections on earnings and the following week its stock is diving. Accelerating complexity and the pressure to respond quickly produce erratic behavior not only in public policy but also among once-steadfast Fortune 500 companies.

Today, fast means competent and slow means inept. Fast means opportunistic and slow means we're falling behind. Fast means nimble and slow means archaic.

In this respect, we are a society that is clear about what we want. We want decisive, action-oriented leaders in the Board Room and in the White House. We want strong, business-minded, rational doers who can quickly assess a situation and act. Never mind that the problems we now face are magnitudes more complex than ever before. Never mind that these leaders possess the same biological apparatus we do and, therefore, are overwhelmed by complexity in the same way we are in our daily lives.

Before we know it, the pressure to act quickly has us comparing Wall Street performance to butter production in Bangladesh, limiting the number of toilets per house, and blaming Saddam Hussein for 9/11. It leads to a society that rushes into war, rushes to write more new legislation, and rushes to fund the next big elixir. Everywhere, rush, rush, rush.

But it is a fact that none of the highly complex, systemic problems that civilization now faces can be solved quickly. Nor can they be remedied easily, cheaply, or by alleviating one symptom after another in a chain of mitigations. If long-term investment and planning continue to be ignored, and we only implement solutions aimed at short-term results, the prospects for a permanent cure are unlikely.

The Fight for a Business Society

To fully appreciate how intrusive the principles of economics have become—to understand the true danger of extreme economics—we turn our attention to centuries-old conflicts in the Middle East.

Like many people, I have ridden the ups and down, the close calls, and near misses of peace in the Middle East since I was a child. Skirmishes would break out; Israel, along with a host of other nations, would offer concessions; a "cease war" would be declared; new factions would then emerge; leaders would convene at Camp David or have tea in Golda Meir's kitchen; and new agreements and alliances would be forged. Then someone would kill a soldier or make a threat and stir the pot all over again. These conflicts began before most of us were born, and from the looks of things will continue long after we're gone.

Though I try, I am ashamed to admit I have a difficult time keeping track of all the different religious sects and their individual beliefs. There's Al Qaeda, Hamas, and Hezbollah. Then there are various tribes that form the Taliban. There's also Abu Nidal Organization, al-Aqsa Martyrs' Brigades, and the PFLP and PFLP-GC, along with the Palestine Liberation Organization, Maktab al-Khidamat, the Egyptian Islamic Jihad, and Palestinian Islamic Jihad. Realistically, I would need a thirty-hour day to study the historical and present-day issues in order to take a responsible position on the Middle East. There are simply *too many* terrorist groups, each with their own history and agendas, and I can no longer keep track of who is doing what to whom.

When knowledge becomes impossible to acquire, then it is customary for us to turn decisions over to a higher authority, and like it or not, this is pretty much what most Americans have done. Aside from academics and a few people immersed in Middle Eastern studies, most of us rely on our government to distill the salient issues and act on our behalf, which, as we found out, doesn't always work.

This is the untenable position complexity causes. We may want to accept personal responsibility, want to take action, but we are paralyzed by our inability to understand all the facts. Eventually, this leaves us no other option but to depend on others to tell us what is right as well as *what is real.*

While recognizing the magnitude of the challenge, and also at the risk of adding to a cumbersome litany of explanations, let me offer a much simpler view of the conflict in the Middle East.

Wars in the Middle East are wars between supermemes.

Some of the disagreements have been the conflict of one religious supermeme against another, but more recently they have become irreconcilable differences between an extreme economics supermeme that dominates Western culture and a religious supermeme that dominates Middle Eastern culture.

Both are entrenched beliefs, neither of which is based on proven facts.

The reason the United States cannot make progress in the Middle East is because we want Muslim nations to make rational decisions based on economic principles. To this end we keep trying to take religion out of the discussion, but they keep dragging it back in. We wonder, "Why won't the fanatics in the Middle East act like mature business people? Why won't they put a better standard of living for their people ahead of religious doctrine?" This seems perfectly reasonable to us. So we keep trying to sanitize negotiations by sidestepping religion as if a business doctrine were not a religion of its own and we were not equally dogmatic missionaries.

In his book *The Stillborn God: Religion, Politics, and the Modern West*, humanities professor at Columbia University Mark Lilla states that "after centuries of strife, the West has learned to separate religions and politics—to establish the legitimacy of its leaders without referring to divine command." Lilla accurately describes the conflict between the Western and Middle Eastern ideology as follows:

> We in the West are disturbed and confused. Though we have our own fundamentalists, we find it incomprehensible that theological ideas still stir up messianic passions, leaving societies in ruin. We had assumed this was no longer possible, that human beings had learned to separate religious questions from political ones, that fanaticism was dead. We were wrong.

He continues:

> A little more than two centuries ago we began to believe that the West was on a one-way track toward modern secular democracy and that other societies, once placed on that track, would inevitably follow.

Lilla describes the U.S. government as a "recent experiment in human history": the first modern attempt to govern independent of a specific religious doctrine. However, his narrow definition of "religious doctrine" is a problem. *Though the West may be trying to separate Christian scripture from governance, in truth it has simply traded one belief system for another: Christianity for extreme economics.*

We only have to look at the West's favorite weapons of choice: trade embargos, tariffs, withholding aid and raw materials, and incentivizing through massive investment in foreign infrastructure and commerce. We bait the traps with our best economic goodies and disincentives and then wonder why nobody bites. And if they do bite once, we wonder why they don't come back for more.

Asking religion-centric societies like those in the Middle East to put business values ahead of their beliefs is actually no different than asking them to think like Christians. The United States may view the pursuit of money as an objective, sanitized process, but in the Middle East they see it as an encroachment on a sacred way of life and are fighting hard to hold on to their own supermeme, even at the expense of a better standard of living. In such an environment, we cannot use economic incentives and disincentives to encourage peace; these will always be dwarfed by the magnitude of the sacrifice mandated by a religion that is actively practiced every minute of every day.

During my travels to the Middle East, I couldn't help but be struck by the many ways religion was integrated into daily life: Grown men in business suits drop a prayer rug on the sidewalk and fall to their knees making customary prayers to Allah throughout the day; loudspeakers encircle the top of every mosque, bellowing religious songs and prayers day and night; women cover their bodies and faces while quietly sitting in Internet cafes surfing the web. Everywhere I looked, every custom I observed was imbued in some way by religious tradition.

Now contrast this against life in the United States, where religious beliefs have been whitewashed. In more recent years, there has been a move toward eliminating Christmas mangers at city hall, silent prayers from public schools, and any mention of God from courtrooms and boardrooms. It's no coincidence that the one place God is still permitted is on our currency. "In God We Trust" appears on every bill, an ever-present reminder to the worshippers of economics to put our faith in the higher power of money.

Proof of the disconnect between the Western doctrine of economics and the Middle Eastern doctrine of Islam came in the 2006 letter that President Mahmoud Ahmadinejad sent to President George W. Bush in the midst of grave political differences with Iran. Imagine the confusion in the White House about how to respond to a head of state who wrote the following:

> I have been told that Your Excellency follows the teachings of Jesus and believes in the divine promise of the rule of the righteous on Earth. . . . According to divine verses, we have all been called upon to worship one God and follow the teachings of divine Prophets. . . . Liberalism and Western-style democracy have not been able to help realize the ideals of humanity. Today, these two concepts have failed. Those with insight can already hear the sounds of the shattering and fall of the ideology and thoughts of the liberal democratic systems. . . . Whether we like it or not, the world is gravitating toward faith in the Almighty and justice and the will of God will prevail over all things.

According to Professor Lilla, when we in the West receive a correspondence like this from another head of state, "we fall mute, like explorers coming upon a ancient inscription written in hieroglyphics." After all, the president of the United States graduated from Harvard's esteemed MBA program and was a self-proclaimed "businessman" determined to bring sound business principles back to the White House. In his view, if governments ran as efficiently and objectively as big business, then many of the problems the world faced would be corrected. The last thing Bush was prepared to do was argue religious history. What would be the point?

What we now have in the Middle East is a complex situation wherein their belief in Islamic scripture may simply be stronger than our belief in the doctrine of economics. Terrorists are willing to risk everything to stop the spread of the extreme economics supermeme to their side of the world because they have seen what business values have done to societies that practice political theology. On the other hand, Western civilization is prepared to do everything necessary, including mounting war, to assure that terrorist nations engage in fair trade, share their natural resources, and behave like good consumers.

Timur Kuran, professor of Economics and Political Science and Gorter Family Professor of Islamic Studies at Duke University, is an expert at "Islamic Economics." According to Kuran, the approach to business, money, and fundamental economic principles is vastly different between Western and Middle Eastern culture. In his article "The Genesis of Islamic Economics," Kuran underscores this point: "More than sixty countries have Islamic banks that claim to offer an interest-free alternative to conventional banking. Invoking religious principles, several countries, among them Pakistan and Iran, have gone so far as to outlaw every form of interest; they are forcing all banks, including foreign subsidiaries, to adopt, at least formally, ostensibly Islamic methods of deposit taking and loan making." Can we in the West imagine outlawing something as fundamental as the right to charge interest on a loan? Ridiculous!

Perhaps no one summarizes the viewpoint of the West better than Ayn Rand's character Francisco d'Anconia in Rand's novel *Atlas Shrugged*, a book that has become a new age bible for business tycoons and economic icons like Alan Greenspan:

"If you ask me to name the proudest distinction of Americans, I would choose—because it contains all the others—the fact that they were the people who created the phrase 'to *make* money.' . . . Americans were the first to understand that wealth has to be created. The words 'to make money' hold the essence of human morality."

The struggle in the Middle East isn't the Christians against the Muslims or the Muslims against the Jews or the Sharks against the Jets. It's more complicated than that. It's the economics supermeme against an equally enmeshed religious supermeme: a crusade of secular economics against a mandate from Allah to stop it.

And when beliefs—whether they are based on economics or religion—overwhelm rational knowledge, there can be no rational outcome.

For the first time in the history of the Middle East, we now understand this dynamic.

Some years ago French author, philosopher, and Nobel Prize winner Albert Camus characterized the role rational thought plays in times of great human turmoil. He said: "In such a world of conflict, a world of victims and executioners, it is the job of *thinking* people not to be on the side of the executioners."

But executioners come in many forms—not only those carrying guns or wielding swords. Some executioners murder desperately needed technologies. Others annihilate tenure and academic freedom. Some destroy the promise and hope of marital bliss with financial contracts, while others eradicate long-range plans and solutions in the name of expediency. Armed with a doctrine of extreme economics, the executioners of human progress come one after the other, as a multitude of victims blindly follow in their footsteps.

Given this scenario, I ask Camus only this: How shall we tell one from the other?

— 9 —

Surmounting the Supermemes

Rational Solutions in an Irrational World

On a breezy afternoon in May in Portland, Oregon, visionary in the sustainable business movement Paul Hawken stood before the graduating class of 2009. Months beforehand, he had been asked to give the annual commencement speech. By his own account, he was determined to give a talk that was "direct, naked, taut, honest, passionate, lean, shivering, startling, and graceful."

Hawken had spent a great deal of his life fighting modern supermemes, so he had a lot to say about the different roles that facts and beliefs play in human progress today:

> When I am asked if I am pessimistic or optimistic about the future, my answer is always the same: If you look at the science about what is happening on earth and aren't pessimistic, you don't understand the data. But if you meet the people who are working to restore this earth and the lives of the poor, and you aren't optimistic, you haven't got a pulse. What I see everywhere in the world are ordinary people willing to confront despair,

power and incalculable odds in order to restore some semblance of grace, justice and beauty to this world.

Hawken has a clear view of our situation. Destructive beliefs that threaten human progress are man made and, therefore, not a permanent condition. So "ordinary people" can surmount entrenched supermemes and once again restore the critical balance between knowledge and beliefs.

Restoring the Balance

There are many historical examples in which individuals, against all odds, triumphed over supermemes. But progress often came at great personal sacrifice. In 1633, Galileo was sentenced by the Catholic church for heresy and ordered to be imprisoned for discovering that the earth revolved around the sun. Though the pleas of his colleagues convinced the authorities to reduce his sentence to house arrest, his movements remained restricted for the remainder of his life.

Charles Darwin was also reluctant to publish *On the Origin of Species* for sixteen years because of attacks he worried his family would suffer from religious literalists. He was right. The church, along with many of his colleagues, friends, and relatives, quickly ostracized him for what they considered to be a heretical theory.

More recently, the persecution, allegations of incompetence, and public humiliation that Martin Fleischmann and Stanley Pons endured forced the two physicists, who claimed to have discovered cold fusion, to seek asylum deep in the English and French countrysides. Yet, since their discovery in 1989, the results of their experiments have been confirmed by so many independent laboratories around the world that science writer for MIT Dr. Eugene Mallove claims the evidence for cold fusion is overwhelmingly compelling and, therefore, a certainty.

Overcoming supermemes is not for the meek.

Thankfully, supermemes such as *irrational opposition, the personalization of blame, counterfeit correlation, silo thinking, and extreme economics* are no match for insightful problem-solving. Dean Kamen, Bill Gates, Jim Watson, Paul Hawken, and E. O. Wilson are not likely to be frustrated by irrational beliefs or common convention. In fact, as the resistance to important discoveries escalates, these unrelenting warriors become more energized and more determined than ever to effect change for the greater good.

For example, in June 2006, financial giant Warren Buffett, worth $42 billion, publicly announced that he would begin giving away 85 percent of his fortune, beginning with roughly $1.5 billion to the Bill & Melinda Gates Foundation, to fight malaria, AIDS, and tuberculosis. Another portion would go to improving education around the world. Then, in 2010, Bill Gates committed an additional $10 billion to develop and deliver free vaccines to impoverished nations. And in 2005 E. O. Wilson, Neil Patterson, and other forward thinkers formed the E. O. Wilson Biodiversity Foundation, which is developing the first comprehensive and free online biology textbook. Their goal is to make everything humankind understands about life on planet earth available to every man, woman, and child.

And there are thousands of other examples.

But by far one of the best illustrations of a pioneer who successfully surmounted modern supermemes is Bangladeshi economist Muhammad Yunus, a man credited with the introduction of "microlending."

Combating Convention

In 2006, Muhammad Yunus won the coveted Nobel Peace Prize for perfecting and popularizing microfinance. For people unfamiliar with Yunus, on the surface he may appear to be an overnight success, but according to Yunus it's been a difficult thirty-year journey fraught with many irrational obstacles along the way.

Yunus began testing the idea that money could be loaned to impoverished people in 1974 in the midst of a great famine in Bangladesh. He started very small. He took 856 taka, which at the time was the equivalent of $27, out of his personal savings and loaned it to a group of forty-two industrious basket weavers who lived in the small town of Jobra. The women used the money to buy materials to sell their wares at the nearby market. They were hopeful that they could earn enough to feed their families.

The story goes that in a remarkably short amount of time, all forty-two women established self-sustaining businesses and repaid the loan, plus interest, in full.

This troubled Yunus.

According to experienced banks and loan sharks, who charged exorbitant interest rates, the poor were risky borrowers. Because they had no assets, the likelihood they would squander the money and not repay the loan should have been extremely high. So when the women repaid his loan quickly and gratefully, Yunus began to question long-held *beliefs* (memes) about lending to the poor.

Were these beliefs correct? Was the threat of losing assets the *only* reason people repaid loans? Or was it also possible that the desire to succeed as a group, or to become self-sustaining, was just as powerful an incentive?

So in 1983, against the advice of financial experts, the government, banks, private lenders—virtually everyone—Yunus formed Grameen Bank (which means "the village bank") based on the principle of "solidarity circles." In addition to requiring no collateral for loans, the new bank's policy was to make loans to *groups* of five to eight individuals. The loans were very small and the duration was short—about six months. In addition, any default of a loan would reflect poorly on *all members of the group,* so when one member fell short, the others were expected to help that person meet his or her obligation.

By using social pressure in lieu of collateral, Yunus unknowingly tapped into a powerful biological phenomenon: our prehistoric instincts to band together and work successfully in small troupes. This is how our ancient ancestors survived in the wild, and it turns out this instinct is still hardwired in our genetic makeup today. We are a species that naturally gravitates toward and depends on a group for survival; anyone who has observed the behavior of our nearest relatives, the

bonobo chimpanzees, can attest to this fact. As a result, when our survival is threatened by extreme poverty and potential starvation, the
predisposition to work together to overcome adversity works just as effectively as the threat of repossessing our homes.

But the proof is in the pudding.

To date, more than 97 percent of Grameen Bank's loans to the
poor have been repaid on time and in full. The bank has made loans
to more than seven million otherwise unqualified and destitute people in fifty-eight countries including the United States, France,
Canada, and seventy-eight thousand villages in Bangladesh.

That's right—a 97 percent repayment rate.

None of the largest, most sophisticated financial institutions in the
world even come close to this track record. What's more, Yunus reports that "since it opened, the bank has given out loans totaling the
equivalent of $6 billion (U.S.). . . . Financially [the bank] is self-
reliant and has not taken donor money since 1995." He continues:
"64 percent of our borrowers who have been with the bank for five
years or more have crossed the poverty line."

Even more important, the success of Grameen Bank has spawned
the growth of thousands of microlending organizations around the
world. According to the article "Small Loans Empower" by microfinance expert Sue Wheat, "Microfinance has now become a crucial
poverty alleviation strategy and there are over 7,000 microfinance institutions worldwide, reaching around 16 million people."

Wheat points out the latest statistics reveal that as many as 1.5 billion people may be living on less than one dollar a day. So organizations such as the Self-Employed Women's Association (SEWA) in
India—which, like Yunus's Grameen Bank, has made short-term
loans as small as $1.50—have become crucial to lifting thousands of
women out of poverty. According to SEWA, up to 94 percent of the
women in India are highly vulnerable financially because they "rarely
own capital or tools of production, and they have no access to modern
technology or facilities." Microlending has become an efficient way to
level the playing field—to give the disadvantaged a fighting chance by
encouraging group collaboration.

Yunus and the Five Goliaths

Microfinancing was successful because Yunus not only tapped into an existing evolutionary instinct common to all humans but also overcame the five supermemes—powerful, entrenched beliefs that inhibit modern progress.

The first supermeme Yunus encountered was massive *opposition* to the idea that uncollateralized loans would be repaid by the poor. According to bankers, loans that were not based on collateral were risky. Yunus also had to confront the long-held belief that lending money to the poor would not make any substantial difference in their condition. Conventional wisdom stated that there were many reasons the poor were poor and that the unavailability of capital was not one of them. Then came the objection to giving groups of people loans instead of holding individuals accountable. To this point, banks made loans only to individuals and businesses, not to ad hoc groups. This was followed by objections to offering loans at low interest rates. Loan sharks had already established that uncollateralized loans would bear higher than normal rates, so why not charge them? Everywhere Yunus turned, yet another objection was waiting.

And even after Yunus finally had hundreds of case studies proving that microlending to the poor worked, he was unable to find financial support to open his bank for another ten years. Supermemes are stubborn and don't easily yield to facts.

Yunus recollects:

You might think that this positive record would cause traditional bankers to change their minds about lending to the poor. But there was not the slightest change. . . . They showed no interest. They had plenty of reasons to explain why the success we'd already enjoyed was sure to end. They could not accept the *fact* that the poor would actually pay back their loans. "The people you are serving must not be really poor," some would say. "Otherwise, how can they afford to repay the loans?"

"Come and visit their homes with me," I would reply. "You'll see that they are definitely poor. They don't even own a stick of furniture! They repay the loans through nothing but hard work, every day."

The second supermeme Yunus overcame was *the personalization of blame.* The prevailing belief among financial institutions was that the poor were poor because of weaknesses in their character—they had bad judgment, were fiscally irresponsible, lazy, or undereducated, and therefore they were underequipped to compete in commerce. These inaccurate perceptions caused financial institutions to view the poor as unfit borrowers who had only themselves to blame for their lot in life. But Yunus didn't see it that way. He believed the poor were simply *people who had no opportunities.* Given the choice, no one would elect to starve to death. So rather than blame the victims for their circumstances, Yunus took a hard look at the systemic forces at work. And once he did, he had an insight: The poor weren't people who had no opportunities. *Opportunities were being systematically withheld from them by erroneous beliefs.* These incorrect assumptions had to be removed before real progress could be made.

The third Goliath Yunus defeated was false data that allegedly proved the poor had a dismal track record of repaying loans: *counterfeit correlation.* The correlation between an absence of assets and loan default was a false one, yet even today this relationship is a widely accepted belief among leading financial institutions. In fact, as mortgage defaults hit an all-time high in the United States, borrowers came under pressure not only to prove they had sufficient assets to justify a loan, but also that they had a regular income. Assets alone weren't enough to offset lending risks—not even if you owned millions of dollars of real estate, stocks, bonds, or a debt-free business. No paycheck suddenly meant no loans.

So it comes as no surprise that it took over a decade for Yunus to amass enough empirical data to overcome a *counterfeit correlation* supermeme that had been bolstered by false data, research, and expert opinions, all proving that making short-term collateral-free loans to the poor was a bad idea.

The fourth supermeme Yunus conquered was *silo thinking.* Yunus

tackled this in two ways. First, he rejected the idea that a financial institution's goals were different from the goals of the community. He
viewed both goals as interdependent: If the customers of the bank pros-
pered, the bank would also do well. Yunus also rejected the notion that
loans to individuals were less risky than loans to groups of citizens. In
his view, operating as individual silos wasn't nearly as effective as collaborating, especially in a competitive and complex environment.

Finally, when it came to the *extreme economics* supermeme, Yunus
put people ahead of profits. In his mind a community that economically thrives is better for a financial institution rather than one that is
victimized by short-term, predatory interest rates. And judging by the
recent consequences of predatory mortgage programs in the United
States, Yunus was correct. In addition to being a successful business,
Grameen Bank could be a long-term instrument of change: The more
profits the bank made, the more loans it could make and the more
people would be affected. It was an upward, never-ending spiral, a
permanent way to reverse poverty. In fact, Yunus's insight was so successful that in 2007 he declared, "Deposits and other resources of
Grameen Bank today amount to 157 percent of all outstanding
loans." Contrast this against the most successful banks in the United
States, which maintain a total reserve of only 3 to 4 percent of their
depositor's money.

By rejecting the idea of *profit at any cost*, Yunus built one of the
fastest growing and most stable financial institutions in the world. He
defeated five powerful misconceptions (supermemes) about lending
to the poor and put human progress back on track again.

If Yunus is a modern-day hero, it is because he has replaced old
myths with proven facts in a field that prides itself on being objective.
But, as he discovered, no industry is immune to supermemes. In this
way, Yunus's insight helped to restore the balance between the impenetrable superstitions that had been perpetuated by the world's most
successful financial institutions and empirical evidence about lending
to the poor. Much like Wag Dodge, the firefighter whose discovery
changed the course of fire safety, Yunus's epiphany—his "aha" moment—changed the course of human suffering.

— 10 —

Awareness and Action

A *Tactical Approach*

Many years ago while I was traveling in Japan, I heard an ancient parable. The story went something like this:

One day a farmer was making his way down a dirt path in the countryside when he had a chance encounter with a Buddha. Noticing that this Buddha had nothing—no shoes, no drink, no food—the farmer took pity on him and offered to share his lunch with the stranger. As they were sitting in the shade of a tree, the farmer couldn't resist asking the Buddha, who appeared in all respects to be an ordinary man, what he was doing walking around the countryside with nothing.

The farmer asked, "Do you think you are God?"

The stranger replied, "No."

"Are you a reincarnation of God?"

"No."

"Are you a man?"

"No."

Exasperated, the farmer exclaimed, "Well then what are you?"

The stranger smiled and replied, "I am awake."

Awakening to the Pattern

To this point we've covered a lot of ground.

For starters, we've discovered that successful civilizations thrive when they vigorously *pursue both beliefs and knowledge* side by side. As evidenced by the Mayan, Roman, and Khmer empires, this occurs during the early years of development, when the population is growing, prospering, and gaining control over its natural environment.

(We didn't know this before.)

Then, as the social systems, governments, rituals, and problems that a civilization must address grow more complex, a society bumps up against a ceiling. Left- and right-brain problem-solving methods, which have evolved over millions of years, stop working. The complexity and magnitude of problems become too overwhelming for the biological capabilities of the brain.

(We've never faced this possibility before.)

The difference between the slow speed at which the human brain can evolve and the rapid rate at which complexity grows is called the "cognitive threshold." Every civilization has encountered the cognitive threshold, and when they did, it marked the beginning of decline.

(Also new information.)

Gridlock is the earliest sign. Leaders and experts become unable to resolve society's greatest threats such as drought, war, and disease and begin passing these problems from one generation to another in an endless inheritance. Individual citizens also begin to feel paralyzed, fearful, and hopeless.

(We never suspected gridlock was part of a pattern. Now it seems obvious.)

Once we reach a gridlock, unproven beliefs take the place of facts and rational thinking. Over time, some of these beliefs become so powerful that they become supermemes. The purpose of supermemes is to compensate for cognitive shortcomings, but they do far more harm than good.

(We never imagined that complexity would lead to abandoning facts.)

Supermemes eventually become so pervasive that they overwhelm all social institutions, customs, values, and rational thinking. Today, five supermemes prevent progress: *irrational opposition, the personalization of blame, counterfeit correlations, silo thinking,* and *extreme economics.* As these supermemes grow stronger, they cause singular ways of behaving and thinking. Singularity in turn suppresses helpful solutions from coming forward. All the while, dangerous problems persist.

(We never suspected everyday beliefs were an obstacle to progress.)

Eventually, one of our systemic problems becomes so colossal that it results in collapse. The final blow can come in many forms: a pandemic virus, global warming, nuclear war. It doesn't matter which unresolved problem goes uncured, one of the threats eventually becomes too massive for the collective resources of a civilization to stop.

Like I said, we've covered a lot of ground.

And that's the good news.

Because understanding is half the cure.

Ancient civilizations didn't have the benefit of examining millions of years of human history. They didn't know about evolution, and they had neither the benefit of neuroscience nor the technology to look inside the human body and brain. Today, however, for the first time in history, this knowledge has converged, permitting modern man to get to the bottom of the biological reasons for decline and collapse.

So the question is: How do we use this knowledge to move forward?

After all, evolution is on its own timetable, so the human brain can adapt only so fast. And from all appearances, complexity is an unstoppable phenomenon that's gaining speed. If the crux of the problem is the discrepancy between two uneven rates of change, is there any way to bridge the gap?

The Short and Long of It

When faced with a seemingly unbeatable opponent—one who is bigger and stronger; one with more money, supporters, or resources; one better armed and also larger in numbers—what do we do?

We do what we always do.

We *outmaneuver.*

Whether it's on the football field or the battlefield, in business or politics, we succeed by acting tactically in the short run and strategically in the long run. In other words, we take action to stop the immediate assault, while we also go to work on attacking the source of the problem.

This is how we win.

For twenty-five years I worked with start-up companies in Silicon Valley, where one of my jobs was to help prepare short- and long-term strategic plans.

A short-term strategic plan meant forecasting products, services, and revenues for the upcoming year, whereas a long-term plan consisted of combing through market projections, competitive data, consumer trends, emerging technologies, and historical patterns until we conjured up a strategy for the next five years.

But more important than the duration of these battle plans was the nature of each. One-year plans were packed with detailed tactics that could be immediately implemented, whereas longer-term plans were much more general. The first was actionable; the other was a telescopic roadmap.

And because product lifecycles in Silicon Valley are short-lived— generally under twelve months before a new technology becomes a commodity or obsolete—one-year plans got all the attention. As a result, three- and five-year plans were reduced to a mandatory corporate exercise for a few weeks every year. Once the process was over, binders filled with elaborate forecasts were put on a shelf and everyone went back to working on what they could accomplish that day, that week, and that month.

So, at the end of grueling planning exercises, I often repeated something that my first boss drummed into me:

> When it comes to planning, keep in mind that our short-term plan is
> to stay in business long enough to *have* a long-term plan.

Twenty-five years later, this same idea applies to our troubles today. *We need a short-term plan to stay alive long enough to get to a permanent cure.*

Fortunately, nature has allowed for both. When it comes to managing complex problems that exceed our cognitive abilities, our short-term plan involves clever mitigations. But mitigations alone are not enough. Thankfully, nature also has a long-term plan: the evolution of *insight*—a growing biological response to complexity.

Mired in Mitigation

As we've seen, the purpose of short-term mitigations (tactics) such as conservation and recycling is to buy civilization the time needed to deploy a lasting solution. However, as problems grow more systemic and there are more wrong answers than right ones, effective mitigation becomes extremely difficult to identify. When we narrowly focus on one or two workarounds, they never impact the problem as much as we had hoped, or as long as we had hoped, even though we often just continue doing them. We convince ourselves that we are taking small "sustainable" measures and that all these measures added together will cumulatively make a difference. But do they?

There are many reasons that short-term mitigations fail.

The first reason is that mitigations often become confused with a cure. And once they are identified as a cure, the hunt for a permanent solution slows to a crawl.

Take vigilante groups now protecting the border between the United States and Mexico, for example. Although their efforts may be successful in the short run, no matter how many vigilantes are recruited to patrol America's borders, it's not a *sustainable* solution to the complex problem of security. Here's why: If sometime in the future the Southwest were to experience a drought similar to the water shortages the Mayans once suffered, mass migration into the United States would be inevitable, and nothing would stop it—not vigilantes, not fences, not ditches, not drones. Imagine millions of people moving northward at the same time in search of water and food. Unless authorities are prepared to order soldiers to shoot women and children illegally crossing the border, the United States will have no way to stop a historic wave of refugees destined to begin migrating as severe climate change takes its toll. In this way, border patrols may be a useful mitigation today when the number of trespassers is small. But as the numbers increase, every measure that is currently under consideration, from walls to canals, is destined to fail.

The same goes for mitigating flulike symptoms at the airports in order to stop the spread of dangerous pandemic viruses. Unfortunately, the symptoms of most viruses don't appear for many days. This means that many human carriers are asymptomatic. As we recently discovered, however, it takes only one tourist, restaurant worker, nurse, or ticket-taker to start a fast-moving epidemic. Again, unless we are prepared to encroach on a person's civil rights by requiring mandatory medical testing and quarantines, there appears to be no way to stop the global spread of powerful new strains of influenza.

Yet, despite acknowledging our superficial stopgap measures, we are a society that regularly confuses mitigation with a permanent solution.

For example, in 2008 Harvard physician Atul Gawande published *The Checklist Manifesto*, a book that received tremendous attention. According to Dr. Gawande, as procedures in hospitals, aviation, finance, and other critical areas become more complex and the opportunity for errors exponentially increases, *checklists* become vital. Checklists significantly lower the number of mistakes and allow everyone, regardless of background and experience, to operate at the same standard.

As with many revelations of this kind, Gawande writes from personal experience. Gawande tells a harrowing story about accidentally cutting a patient's artery during surgery. Normally, this would have caused blood to fill the patient's body cavity and the patient would have died on the operating table from rapid blood loss. But in Gawande's case, the patient survived thanks to an operating room checklist that required four units of blood to always be on standby.

Gawande applauds the use of checklists in other highly complex fields as well. In his book he reveals that checklists have been used for decades in aviation to prevent air disasters. He cites many examples, from finance to cooking to construction, where highly complex tasks have been broken into simple checklists, thereby resulting in greater consistency and productivity.

Gawande also explains how checklists can be used as an effective antidote against mounting complexity in our personal lives. He says that "checklists seem able to defend anyone, even the experienced, against failure in many more tasks than we realized. They provide a kind of cognitive net. They catch mental flaws."

Although Gawande makes a compelling argument for the use of checklists, he also seems unaware that he is advocating a temporary mitigation, which, like other mitigations, *isn't sustainable.*

As complexity continues to grow, the number of items on a checklist would also grow, and given the current acceleration, grow rapidly. After all, the more we know about what can go wrong, the longer a checklist becomes. A ten-item list quickly grows to twenty items, then from twenty to fifty, and from fifty to one hundred until eventually the list is too long and has to be broken into many smaller checklists to be useful. It doesn't matter if we have one list with a hundred items or ten lists with ten items each, as tasks become more complex, they also become unmanageable.

That's because the root of the problem isn't organization but rather complexity. Operating room procedures are different for every type of specialized surgery, and every year the number of safeguards needed exponentially grows. The same is true with today's highly complex aircraft, weaponry, finance, and information systems.

It doesn't take long before checklists start looking a lot like other short-term mitigations such as recycling, mailing stimulus checks, or riding in jeeps along the border to stop illegal immigration. Although each of these measures temporarily ameliorates the symptoms of a complex problem, they do little to address core issues.

This brings us to the second reason mitigations fail: *Mitigations remove the sense of urgency to solve our problems.*

Mitigations give us the impression that the problem is solved when symptoms begin to recede. Once they recede, it is human nature to become lackadaisical about continuing to work on the problem—especially when a longer-term solution is more expensive and more difficult, and its impact may not be felt for years. So, instead of putting our efforts toward finding a permanent solution, we are much more inclined to implement one temporary fix after another.

As an example, once the U.S. government had bailed out the American automakers and averted the loss of millions of jobs, all mention of the automobile industry disappeared from the news. Likewise, once Bernie Madoff was handcuffed, the nation also breathed a sigh of relief. The system had worked, and wealthy Wall Street crooks were surely on the run. And once everyone in California stopped watering their lawns, the worrisome drought suddenly seemed manageable.

Even though logically we know the problem hasn't been fixed, once the pain subsides, we simply lose our appetite—our motivation—to tackle tougher systemic problems.

The third reason mitigations don't work is that individual mitigations prevent systemic problems from being solved in a systemic way.

The most threatening problems we face today are all systemic, and systemic problems require highly complex, multifaceted solutions. These problems cannot be reduced to simple cause and effect, and therefore solutions based on simple cause and effect don't work. Translated, this simply means that the cure to our biggest problems requires money, energy, focus, endurance, and the postponement of gratification for unbearably long periods of time. And who would put up with that?

On the other hand, mitigations by their very nature are cheaper, easier, faster, and produce an immediate result. But every mitigation we can name—financial bailouts, water rationing, cease-fires, border patrols, and increased airport security—addresses only one or two aspects of a larger problem. So, in the end there's never enough critical mass to effect systemic change.

The fourth problem with mitigations is they are not sustainable.

As we have seen, complexity isn't static. It's a moving target that grows and accelerates over time. This causes mitigations to be effective for shorter and shorter amounts of time. So, as each mitigation fails, we rush to replace it with Plans B, C, D, E, and so on.

And finally, mitigations fail because we take a serial approach to improvement—one finger in the dike at a time despite there being ten holes.

Mitigation That Matters

When the problems we face are systemic, rifle shots don't work.

But Gatling guns do.

Single mitigations may not have the power to slow the momentum, but when many mitigations are implemented in parallel, their sum quickly exceeds their parts.

One of the finest examples of the effectiveness of parallel mitigations occurred in the United States during World War II.

By the time the United States entered World War II, twenty-two countries spanning the Asian, European, and African continents had already been embroiled in a global conflict for over two years. Even as the conflict continued to escalate, the United States insisted on remaining a neutral nation—that is, until December 7, 1941, following the Japanese attack on Pearl Harbor. One direct attack changed everything. The day following the assault, the United States quickly entered the conflict—*but only against Japan.* It wasn't until December 11—after the United States officially declared war on Japan—that Germany and Italy declared war on the United States. So in the short span of a few weeks, a country that had fought hard to stay out of global turmoil suddenly found itself smack-dab in the middle of a struggle for power the likes of which humanity had never experienced. Vying for domination of the entire planet, Germany, Italy, and Japan along with Hungary, Romania, and Bulgaria squared off on one side against the United Kingdom, the Soviet Union, the United States, and 50 other Allies on the other.

With so many battles taking place at the same time, in so many different parts of the world, and so many different armies with different leaders, weapons, protocols, and training involved, there was only one thing the United States could do: Gather the Allies together and launch as many tactics as possible *all at once*.

In addition to expertly coordinated logistics by the military commanders of the three strongest Allies (Stalin, Roosevelt, and Churchill in the European Theater and Chiang Kai-shek, Roosevelt, and Churchill in the Pacific Theater), every institution, man, woman, and child in the United States rallied. Women left their homes and went to work in factories, building planes and packing ammunition; domestic food and fuel rationing was enthusiastically embraced; schools installed bombing sirens and mandated practice drills; young men received permission to leave universities to enlist in the service; and fully credentialed doctors and nurses signed on in record numbers as military medics. Consumers began shunning foreign-made products and began buying government bonds in greater numbers than ever before to support the war effort, while airports opened USO facilities. The government enacted the GI Bill to help veterans reassimilate back into the economy upon their return, and immigration restrictions were eased so as to allow record numbers of war refugees to easily enter the country. On any given day a parade welcoming returning soldiers could be found somewhere in America.

During World War II, every institution and every individual in every allied nation were willing to do their part, but the key was that they all acted *in unison*. So, it didn't take long before thousands of small and large incremental efforts added up to one colossal victory. Germany, Japan, and Italy were defeated, and once these countries were defeated, the leaders of the Allied forces spared no time in going to work on longer-term measures in order to make certain the world never faced a similar conflict again. They had the wisdom not to stop working on a permanent solution just because the fighting was over and mitigations had succeeded. Instead, they persevered, optimistic they would find a permanent way to put war to bed once and for all. In the aftermath of World War II, the United Nations was founded, NATO was formed, all industrialized nations adopted the treaties of the Geneva Conventions, and countless other innovative measures were undertaken to resolve what was clearly a multifaceted problem.

From both a short- and long-term perspective—a mitigation as well as a cure—World War II marked a high point in our ability to understand, manage, and address a highly complex global problem. Humanity triumphed against an unimaginable challenge and then wisely put safeguards in place to ensure sustainable peace.

Many people who lived through the war look back with nostalgia at this period. They feel they were part of something big, something important, because everyone was focused on a single objective. Some people call this patriotism, but in reality it's a heartwarming example of the power of *parallel incrementalism*.

Parallel incrementalism is a mitigation strategy whereby the cumulative effect of executing multiple, incrementally useful mitigations in tandem is exponentially more effective than implementing them one at a time.

In simple terms, there comes a time when a problem is so big and so complicated that you have to throw everything—including, perhaps, the kitchen sink—at it.

If It's Not One Thing, It's Another

Parallel incrementalism doesn't apply only to periods of war or natural disasters. We also use it in our private lives.

Say, for example, we, like millions of citizens, suddenly found ourselves in danger of losing our home because of rapidly rising mortgage rates. What options would we have?

Well, the first thing we would do is stop unnecessary spending. We would trim our budgets to the bare bone and then trim again. We would probably also cash in our 401k, stocks, bonds, and other investments. We might also try to refinance our loan at a lower rate and contact our credit union and visit our bank to see how they could help us. We would apply for every available government relief program and reach out to one of a dozen credit counseling companies for advice. We would sell our second car, have a garage sale, take our jewelry to a pawnshop, and look for a second or third job. Some of us would also rely on family members and friends for financial help.

In other words, we would do everything and anything we could to save our family home. And we would do all of these things at the same time—not one at a time.

Some of the things we tried would have a big impact, and some would produce no result at all. For example, maybe we don't meet the qualifications for FDIC assistance, but instead we discover our credit union can help us. Or maybe the garage sale and second car didn't bring in as much cash as we hoped, but the jewelry we pawned was worth a lot more than we thought.

The point is that under fire, the best strategy is to *try everything*. We can't wait for Plan A to fail before we initiate Plan B; this approach takes too much time and our home would likely belong to the bank by the time we got around to our last idea. Instead, when times get tough, we have to do our best to launch Plans A, B, C, D, E, and F all at once, hoping that a few of these ideas work, even if we don't know exactly which ones.

This is parallel incrementalism.

Parallel incrementalism assumes that the sum of our incremental effort offers us the greatest chance of success. The more complex and destructive a problem is, the more critical it becomes to launch mitigations in *parallel*. That's because when faced with excessive complexity it's impossible to tell beforehand which solutions will work. In fact, sometimes it isn't even possible to prioritize our options.

Consequently, parallel incrementalism is the most effective response when we are about to lose our home, our life, or our freedom. It also happens to be the most effective response when humankind is under threat. Whether it's trying to solve a catastrophic oil spill, the dangerous spread of waterborne diseases, or a global recession, casting a wide net is one way to counteract complexity.

Wisdom in Venture Capital

Once, while being interviewed, Thomas Edison is said to have remarked, "I have not failed. I've just discovered 10,000 ways that don't work!"

It's interesting that when it comes to inventors, engineers, scientists, and explorers, no one is surprised to learn that the vast majority of their attempts fail. Their work often produces no tangible results for long periods of time—and sometimes never. In fact, the more difficult and complex the problem they are attempting to solve is, the greater is the failure rate we have come to expect.

Yet, if you ask any person on the street whether they would elect a president, senator, governor, or mayor who failed 80 percent of the time, it would take less than a second for them to answer, "No." How about the CEO of a large public company, your financial advisor, or your doctor?

Probably not.

But there's a very successful business model that is ideal for complex problems because it is built on failing the vast majority of the time. It's called venture capital.

Venture capitalists are experts at failure. These adept individuals fund thousands of new companies in speculative, unproven markets each year, all the while acknowledging that less than a quarter of their investments will be responsible for all of their profits. Though venture capitalists hope that every investment they make will be profitable, the reality is that about 20 percent of their portfolio companies return over ten times their initial investment. About 50 percent of the companies return a very modest profit, and the remaining third of the companies are worth less than the money invested. Nonetheless, the profits produced by the 20 percent that are successful make the failures look miniscule. In other words, an extremely high rate of failure is simply the price tag venture capitalists are willing to pay for hitting the jackpot on a small number of successes.

Curiously, no one accuses this business model of being flawed.

But why not?

Trying to predict how an unknown technology will fare in an unknown market is so complex, so unpredictable that no amount of additional due diligence can better the odds. Therefore, it can be said that venture capitalists are *in the business of making many more wrong calls than right ones.* The key is to make *enough* calls so that the successes have an opportunity to neutralize the losses.

Another way to say this is that successful venture capitalists are experts at managing high levels of uncertainty and complexity because their business model is based on *parallel incrementalism*—many incremental investments made in parallel.

However, imagine if the venture capital community became paralyzed by their 80 percent failure rate. Imagine if they suddenly found themselves investing in fewer and fewer companies in a futile attempt to lower their risk. Imagine if they began holding a start-up, any start-up, to the same standards to which we now hold our political leaders or the heads of AIG or Ford Motor Company.

Their business model would quickly collapse.

This is what happens when a society becomes intolerant of wasted efforts, wasted resources, wasted time, wasted research, and wasted optimism. When we demand that every proposed solution and program be successful *before* we invest a nickel in it, progress comes to a halt.

Ironically, as our unsolved threats grow in magnitude and complexity, our intolerance of waste also grows. The more desperate our situation becomes, the more exacting a solution we demand—and the more precision we require from our leaders. Eventually, the bar for guaranteed success is set so high that no proposals can be adopted or supported. We oppose everything on the grounds that it is flawed, unproven, or inefficient, and this exacerbates the gridlock we find ourselves in.

Take the recent global recession as an example. Just a few months after the U.S. government bailed out financial institutions on the verge of bankruptcy, we were already complaining about the wastefulness of the AIG bonuses. Then we zeroed in on the wastefulness of auto executives who flew their private planes to Washington. In our minds, the bonuses and private planes meant irresponsible spending.

Our tolerance for waste and failure is dangerously low. If we had been thinking like a good venture capitalist, we wouldn't have batted an eye if 70 percent of the bailout money was wasted and only 30 percent produced the turnaround intended by the government. What if only three out of every ten dollars invested stopped American automakers from sinking into a deeper recessionary spiral? Is there any doubt that we would consider that to be a colossal failure?

But when a civilization is faced with indiscernible complexity, progress is dependent on the amount of wasted effort and resources the civilization is willing to tolerate.

In other words, the more complex problems become, the less efficiency we must expect. When we can't understand the cause or nature of our greatest threats, then there's very little chance we can be accurate about which cures will work. As a result, we have to be willing to accept as much waste as a venture capitalist in order to keep moving forward. To derive benefits from a small number of programs that have big payoffs, we have to accept the fact that in a highly complex environment more solutions will fail than will succeed. This is the price of progress.

Parallel Plans in Public Policy

We understand how parallel incrementalism works in venture capital, but how does that translate to public policy?

That's easy. Instead of jumping from one failed mitigation to the next, we simultaneously launch every solution that can incrementally improve the situation. In other words, the way to attack our largest, most threatening, and most persistent problems is to use everything and anything we've got: No more wait and see if it works, no more one-shots, no more further study to determine which is the best bang for the buck, and no more arguing about which approach is best.

For instance, in a perfect world we would proactively initiate multiple public works programs to neutralize the effect of the coming drought—and do it today: build desalination plants, construct massive new reservoirs, offer tax incentives for building private underground cisterns, deploy new technologies designed to extract water from the atmosphere, build massive pipelines for transporting water to arid areas, invest in research aimed at "cloud seeding," and so on. We wouldn't argue over *which of these* programs was the most useful or cost effective. And we wouldn't rely on serial mitigations because we'd recognize that time would run out before we solved the problem.

But sadly, the areas that will be most affected by a drought in the coming years are doing very little to prepare for it. Like the Mayans, who had thousands of years to implement a plethora of practical remedies, modern society is succumbing to irrational denial and convenient beliefs.

In this respect, California is currently managing its dangerous water shortages the same way world governments address other dangerous threats. Take a look at overfishing, for example.

It is a widely known fact that humankind is depleting ocean life faster than the ocean can replenish it. As technology makes it easier for large fleets to locate and catch record yields of fish, we are having a devastating impact on a critical food source as well as an entire marine ecosystem.

Though the statistics widely vary, they are all so alarming it doesn't really matter which figures you are inclined to accept. According to Nick Nuttall of the United Nations Environment Program (UNEP), the Food and Agriculture Organization (FAO) reports that as of 2004 "over 70% of the world's fish species are either fully exploited or depleted." Nuttall reports, "In the last decade, in the North Atlantic region commercial fish populations of cod, hake, haddock and flounder have fallen by as much as 95%, prompting calls for urgent measures."

The consequences of overfishing our oceans are only now beginning to surface. For example, the overfishing of anchovies along the Peruvian coast resulted in yields dropping in just one year from 10.2 metric tons to 4 metric tons. And in 1992, cod fisheries off of Newfoundland, Canada, completely collapsed, leading to the loss of approximately forty thousand jobs. Fisheries in the North Sea and Baltic Sea are now quickly following suit. Greenpeace laments, "Instead of trying to find a long-term solution to these problems, the fishing industry's eyes are turning toward the Pacific—but this is not the answer."

It seems like no matter where we turn, short-term mitigations, such as fishing a different ocean, run rampant. But as long as the problem is not in our backyard, as was also the case with building more prisons, what do we care?

The overfishing of our oceans is a highly complex, systemic problem that involves many, many factors. First, there are the rights of the commercial fisherman to consider. Many have fished the same waters, and earned a living doing so, for multiple generations. Can anyone blame them for leveraging the best sonar equipment and nets to capture the highest yields possible? Then there are the fish—what chance do they have against technology? It isn't as if they can evolve new evolutionary tools to outmaneuver man's growing competency. Consider also the consumers. Fish are a major source of protein for a large part of the world's population, so overfishing has a dangerous impact on many people's ability to survive. What's more, we must also consider diplomacy with other countries, as fishing boats begin to cross over long-respected borders in the ocean just to fulfill their quotas. Lastly, of course, there are the ecologists, biologists, naturalists, academics, and their followers, who are concerned with the untold damage overfishing is doing to marine life and the planet as a whole.

It's clear that fishing involves the rights and concerns of many groups. It's a highly complex, multifaceted issue—one not likely to be successfully mitigated by creating "sustainable fish" restaurants or raising the price of cans of tuna at the grocery store. It's not that simple.

In this respect, stopping overfishing isn't really that different from trying to resolve terrorism, overcrowded prisons, or a worldwide financial meltdown. It's no different from the unilateral effort required to fix a failing public education system or stopping the pandemic of addiction, and no different from the challenges we once faced during World War II. In each instance, successful mitigation requires attacking the problem along multiple fronts: politically, economically, legally, internationally, educationally, culturally, and ecologically. Until every aspect of the overfishing problem—from the rights of the fishermen to make a living, to the rights of people who put the planet ahead of profits—are addressed with equal fervor, the problem will not be stopped.

Parallel incrementalism succeeds because it forces mitigations to address the systemic issue rather than treat individual symptoms.

The Long Term

Parallel incrementalism is the most effective way to leverage mitigations when problems become too complex to be understood or managed. However, any mitigation strategy—whether single or multifaceted—is temporary. What about more permanent solutions? Can we do anything to permanently close the gap between complexity and cognition?

Here again, modern man has two weapons earlier civilizations didn't: the restoration of knowledge and the evolution of insight.

The first antidote to collapse is to *restore the balance between knowledge and beliefs.*

One of the ways we break the historic pattern of human ascension and decline is to vigilantly maintain a balance between a civilization's pursuit of beliefs and facts. As we discussed earlier, societies and individuals thrive when unproven beliefs and proven facts exist side by side. It is only when complexity causes the acquisition of knowledge to become too difficult that beliefs take over. And as we have seen, over time, some of these beliefs become dominant supermemes that obstruct human progress.

To that end, when a society elevates the pursuit of knowledge, it counteracts the encroachment of beliefs and, in effect, inoculates a civilization against supermemes. Think of facts and knowledge as powerful vaccines that protect us from surrendering to irrational beliefs.

There are many ways modern civilization can protect the pursuit of knowledge.

One way is to preserve the integrity of educational institutions against the influence of supermemes. As we discovered earlier, universities and colleges are now increasingly beholden to large businesses for funding. Consequently, research that demonstrates commercial viability is more likely to receive funding than research for knowledge's sake, and this has a dramatic effect on a university's priorities.

Our important learning institutions represent a barometer of how the relationship between beliefs and knowledge has become imbalanced. The more supermemes (such as profitability, productivity, return on investment, and other tenets of economics) drive the agendas of higher learning institutions, the less likely it is that universities will engage in research that has no short-term, leveragable, or tangible benefit. Many university administrators admit they are dismayed by this problem yet feel compelled to go along to prevent financial ruin. They're trapped—trying to answer to the principles of economics while also trying to maintain reverence for the pursuit of knowledge and advancement of humankind.

Today's universities would do well to proactively set a limit on how much of their total contributions are allowed to come from business. Rather than take money where they can get it, they would be wise to resist the temptation to become the unintended research arm of commercial enterprise.

Although universities are an obvious place to start, there are many other arenas where knowledge is also losing ground.

For instance, the budget and time required for investigative journalism are quickly becoming a thing of the past. It is rare for a newspaper or news program to commit the resources needed for any in-depth, long-term research project. No one can afford it anymore. The consequences of rushing to get a story out without requiring second-and third-party sources means nothing short of printing hearsay—just ask news anchor Dan Rather, who erroneously reported on President George Bush's military record, or Jayson Blair, the *New York Times* writer who was accused of plagiarizing and fabricating facts for his stories between 1999 and 2003.

Even in our private lives, we have very little time to pursue knowledge. Complexity has not only made facts difficult to discern; the Internet and fast communications have made them too plentiful. So much is coming at us so fast that we can't tell left from right. In such an environment we have very little choice but to rely on experts, celebrities, and talk show hosts to screen the relevant "facts" for us and draw conclusions on our behalf. So although it's tempting to point the finger at universities, the media, or other institutions, we also have to accept some of the responsibility for being unwilling to put in the time and effort needed to separate fact from fiction.

There are many ways to elevate rational thinking, research, and scientific proof so as to guard against paralyzing beliefs. It might require more fact-checkers in the newsroom or dedicating the time and resources required for truly in-depth journalism. It may call for the government to subsidize newspapers so that everyone has access to one at no charge. It may require a society to make teaching one of its highest paying jobs in order to draw the best and the brightest to the field—professionals who can imbue first-graders with the joy of learning and inspire young adults to make a clear distinction between what they believe is true from what can be proven.

In a better world, elevating the role of knowledge would lead to friends and neighbors hosting weekly "discussion groups" designed to get to the bottom of confusing topics ranging from health care and investing, to global warming and terrorism. Perhaps these discussion groups would become as popular as weekly book clubs, which, thanks to Oprah, once swept the country. Perhaps people would insist on reading and listening to more than one perspective. Followers of Fox News would watch MSNBC once or twice a week and vice versa. One of my good friends reads two newspapers every morning—one domestic and one foreign—just to get as objective a vantage point as possible. She admits it's cumbersome and costly, but she also claims it is the only way she's found to avoid cultural biases.

When we elevate the pursuit of knowledge to be equal to the pursuit of money and convenience, civilization will rediscover the balance it needs to thrive. This means knowledge must become as desirable as celebrity, as valuable as winning, as commonplace as opinion, and as revered as tradition. Then and only then will knowledge and beliefs coexist side by side.

And then and only then will the tired age-old debate between religion and evolution come to its proper end. As we acknowledge the human need for both beliefs and knowledge, we arm ourselves with a new weapon against extremists who demand that we choose one over the other. We may not understand precisely how evolution fits with religious doctrine today, but that doesn't mean one of them is wrong and therefore unnecessary to human progress. It simply means we do not have the cognitive ability to reconcile the two—like Einstein's theory of relativity and Newton's theory of gravity, we haven't figured out exactly how the two work together yet. But so what?

As long as we continue to insist that one idea is right and the other wrong, the balance between knowledge and beliefs remains in jeopardy. Whether it is scientist and atheist Richard Dawkins, who insists that religious beliefs do humanity a great disservice, or men of the cloth who frequently stand in the way of rational thought, both do harm to the progress of civilization by ignoring historic evidence that we are a species that has, from the beginning of time, pursued data and divinity. Experts on both sides of the argument ask nothing less of us than to choose between two equally precious children: our need to know and our imperative to believe.

Watchdogs in the White House

There is no place where knowledge and fact are needed more today than in our nation's leadership. As the complexity of dangerous global problems mounts and the pressure grows to solve problems faster and better, government leaders find themselves more often basing their decisions on ill-conceived logic and unproven beliefs than on rational facts. The confusion over what is scientifically sound data and what is merely theory, correlation, or guesswork causes leaders at the highest levels to jump to dangerous conclusions—leading entire nations astray.

For example, in hindsight we can now admit that there was very little tangible evidence for the existence of weapons of mass destruction in Iraq, even though this was the basis for the U.S. invasion in 2003. Colin Powell presented a few fuzzy satellite pictures on the floor of the United Nations, and there was some on-the-ground intelligence that could not be confirmed. That, in a nutshell, sums up what the United States knew.

Under a heightened sense of danger following 9/11, U.S. leaders became more confused than ever about the difference between facts and beliefs. With their desire to crush the enemy and with very few facts to go on, in 2002 the House and Senate voted in unison "to authorize the use of United States Armed Forces against Iraq." In this way, the confusion between what was real and what was speculation was responsible for a hasty and irrational decision that cost thousands of military and civilian lives.

Sadly, war is only one glaring example in which beliefs grow so strong that leaders and experts begin confusing them with evidence. More recently, the argument over health care reform in the United States was largely waged by politicians with law and business degrees. As far as I know, there was only one doctor among them. These same lawyers are also responsible for establishing policy on other complex issues such as stem cell research, global warming, nuclear technology, and education. Armed with reports and data they are not trained to understand and that are too voluminous to read or carefully examine, they are forced to rely on instincts, staff, politics, and beliefs. No one expects a government official to be an expert in every field, but neither do we expect them to make superficial rulings concerning dangerous issues without consulting the most credible authorities in the field.

But what happens when they don't? What happens when there simply isn't enough time or money to bring in experts or sort through a never-ending mountain of data and informed opinions? Is there any way to safeguard against making policy decisions based on beliefs?

It turns out there is.

Few people know that in 1951 President Harry Truman shared these same concerns. So Truman began discreetly inviting scientists to the White House to meet privately with him. The scientists were never paid, nor were they seeking fame or fortune or a permanent job. Truman made it clear that the purpose of the meetings was for the president to get unvarnished facts directly from the mouths of the brightest minds in the nation. This was Truman's way of balancing the truth against political agendas that might otherwise get in the way of making the best decisions for the country.

Then following Truman, President Dwight D. Eisenhower also called upon the nation's preeminent scientists for counsel. After

World War II, the United States grew increasingly fearful that it was losing its leadership in science and technology. So in 1953, Eisenhower began informally inviting small groups of scientists to the White House. Eisenhower claimed that this was the best way to counterbalance information he was receiving from cabinet members and the U.S. military regarding emerging radar, nuclear, and space technologies.

By 1957 with the Russian launch of Sputnik, tensions in the Cold War reached a new height. Sputnik was all the proof the military needed to conclude that "the Soviet Union was dangerously ahead on intercontinental missiles."

It was during this worrisome time that President Eisenhower found himself relying more and more on experts outside of Washington. In fact, the scientists' input became so indispensable that the president decided to formalize the organization, renaming Truman's original gatherings the President's Science Advisory Committee, or PSAC.

Later, President John F. Kennedy perfected the role PSAC played in making complex scientific research discernable so it could be used to shape public policy. Kennedy also broadened PSAC's predominantly military focus to include space exploration (NASA) following his commitment to put the first man on the moon.

According to Charles Townes, Nobel Laureate in physics and frequent PSAC participant during the Eisenhower and Kennedy administrations, fifteen to twenty of the most knowledgeable scientists in the United States traveled to the White House once a month to discuss urgent topics with the president. The scientists rotated in and out according to their expertise. Aside from the chairman, there was no permanent paid staff. The objective of PSAC was to give the nation the opportunity to benefit from the greatest minds in the country a few days each month. It was a civic duty that every scientist and expert welcomed.

When the scientists arrived in Washington, they were quarantined for one to two days to confer among themselves with no supervision or interference. Then, on the last day, they met privately with the president to deliver the results of their deliberations in a frank, uncensored exchange. There wasn't always agreement among the experts nor with the president himself. But once they presented their findings—win, lose, or draw—the temporary advisors returned home to their own work.

Then, in 1973, for unstated reasons, President Richard Nixon disbanded PSAC.

With this action, a critical conduit for objective facts was lost. The termination of PSAC was no less a sign that the United States was headed in the wrong direction than the Mayans' decision to stop building reservoirs and begin appeasing the gods through human sacrifice.

Richard Garwin, author of *How the Mighty Have Fallen*, points to the fact that independent experts have been "increasingly subject to political and ideological tests," which has made it difficult for objectivity to permeate the Oval Office. He notes that many people view the abolition of agencies who deliver the unbiased facts—agencies such as the Office of Technology Assessment and PSAC—as being equivalent to "shooting [ourselves] in the brain."

That said, in 2001, President George Bush attempted to resurrect a PSAC-like organization by forming the President's Council on Science and Technology (PCAST)—an organization comprised of presidential appointees. According to Bush, PCAST represented an improved version of Eisenhower's original organization. But PCAST bears only a slight resemblance to PSAC—a once vital source for "straight talk" to the president.

For example, PCAST now operates under the close supervision of the Office of Science and Technology (OSTP). OSTP is a department comprised of full-time Washington employees managed by Dr. John Holdren, who not only serves as the co-chair of PCAST but is also the Director of OSTP and the Assistant to the President for Science and Technology. This means that all of the experts, appointees, research, recommendations, etc., are now scrutinized by Holdren and OSTP staffers prior to reaching the president.

But the original purpose of PSAC was to counterbalance the opinions of those *inside* the Washington beltway regardless of the political fallout or whether the scientists' recommendations supported the existing administration's policy. This was also one of the reasons all PSAC meetings were held privately with the president with few staffers in attendance and no media or public access. This allowed the scientists to speak freely without having to worry about the administration's agenda or the safety of their jobs. Contrast this against the PCAST meetings today: They are managed ostensibly by Holdren and OSTP, and each proceeding is open to media coverage, which, under the pressure of public scrutiny, inhibits the free exchange of new ideas.

Take the recent Gulf oil spill, for example. A fixed panel of 35 presidential appointees met to make recommendations—none of whom had any specific credentials related to the failure of offshore oil wells—yet scientists who had immediate solutions to one of the greatest ecological disasters in U.S. history found no seat at the table. Rather than reach outside the White House walls to assemble a highly specialized team of experts, the president predominantly relied on the office of OSTP and PCAST appointees for several days before turning to outside counsel.

In fact, things moved so slowly in Washington that private citizens attempting to get through to OSTP, the White House, or any other agencies—people such as actor and advocate Kevin Costner—had no choice but to personally take action by deploying their own army of experts, all the while urging other groups *not to wait for Washington* to implement technologies readily available.

So, at the risk of being accused of misinformation or exhibiting oppositional behavior, allow me to be the first to publicly say it: PCAST is not PSAC. The once-fluid group of specialists who assembled monthly to deliver the unvarnished truth to the highest office in the nation has slowly been reduced to an extension of OSTP—far from the objective and powerful problem-solving entity it once was.

Yet, at no time has the public and planet needed great minds to come forward more than we do today.

As modern society approaches the cognitive threshold, there is no better time to restructure PCAST or resurrect the President's Science

Advisory Committee and to encourage every representative and senator to meet with experts outside of the Washington beltway as often as possible. Why not make this the new role model for the world: a government that leans on a small army of unbiased, forward thinkers to defend itself from the encroachment of supermemes, a government that uses its intellectual capital to triumph over the ensuing gridlock and pave a rational and sustainable course? If the United States believes its rightful place in history is to lead the free world, it must now lead by example. In this way, the true gift the United States offers humanity is safe passage across an inevitable and dangerous cognitive threshold.

In the End, Evolution

Remember the ancient Japanese parable? When the monk was asked whether he was God, a reincarnation of God, or a man, he simply answered that he was "awake."

What did he really mean?

He meant that he could see what had been previously hidden from view. And with clarity of vision came clarity of purpose and action.

The same can be said for humankind today. We now have clarity. And so we can act.

When it comes to tackling complexity, we have the ability to evoke parallel incrementalism and restore the critical balance between knowledge and beliefs. This immediately buys us time. But that's not all. We also have a permanent solution that comes from none other than Mother Nature herself: *insight.*

It is an undeniable fact that no matter what challenges a biological organism encounters, one way or another evolution gets around to solving them. Over 150 years ago, Charles Darwin demonstrated that this is one principle we can count on. So from the perspective of responding to our environment, our changing cognitive abilities are really no different than adopting two-legged locomotion. At any moment in time we are improving our opportunities to survive by biologically adapting to our environment—and in the case of modern man, complexity is a powerful driving force.

And though the evolution of insight may be painfully slow, there is little argument that, in the long run, it is the Holy Grail: nature's elegant defense.

Thankfully, today's neuroscientists are on the cusp of harnessing this shy, emerging talent in the human brain.

And with little time to spare.

— 11 —

Bridging the Gap

Building Better Brains

One day when I was still in grammar school, a teacher told me that I was only using 10 percent of my brain.

This really bothered me.

Of course, the teacher who mentioned this neglected to say that all of us were using 10 percent. So, naturally, I assumed it was just me.

After a week of feeling bad about her remark, I went to my dad—the smartest person I knew. Dad had an answer for everything. Flat tire on my bike? Here's a patch kit—read the directions. Some kid was mean on the playground? Play somewhere else. Can't get my chores done? Get up earlier.

One night I popped the question out of thin air:

"How can I use more of my brain?"

"What?"

"I want to know how to use more of my brain."

He put his newspaper down. "You study at school, that's how."

"But I already do that and my teacher said I'm only using 10 percent of it."

"That's right. That's because that's all you can use. Everybody uses 10 percent."

"But I want to use more of it."

"Well that's easy..." He paused and drew a deep breath, "Once you think you have the answer.... Once you're really sure you know what it is and you're positive you've got the right answer, you force yourself to come up with another one."

"Another answer? That's it?"

He continued, "Well, now wait a minute. Pretty soon you'll be like a runner who runs a little farther every day. Before you know it, you can run a marathon. After that, you can run longer and faster than anyone's ever run. If you want to use more than 10 percent of your brain, here's what you do: First, you leave your parents alone when they're trying to read the paper; then, second, start working on using 11 percent."

Powerful words from a simple man.

I LEARNED MUCH LATER that the 10 percent rule is a myth.

It's a stubborn meme that's been around since the late 1800s. The story goes that psychologist William James was studying the physiological limitations of athletes when he became interested in a phenomenon they called "a second wind." From his research, James extrapolated that in addition to rarely using all of our physical potential, we also rarely exert all of our mental potential. From here, it was a short step to assuming that most of the time we only use part of our brains. Then a group of researchers reported that "a huge percentage of the cerebral cortex was the *silent cortex*." Once scientists admitted that a large part of the brain was "silent," the public incorrectly interpreted this to mean this part of the brain was *doing nothing*.

But it wasn't true. *We just didn't know what the rest of the brain was doing.*

As Dr. Barry L. Beyerstein, professor of psychology, now deceased, at the Brain Behavior Laboratory at Simon Fraser University in Vancouver, explained in his article titled "Do we really use only 10 percent of our brains?":

First of all, it is obvious that the brain, like all our other organs, has been shaped by natural selection. Brain tissue is metabolically expensive both to grow and to run, and it strains credulity to think that evolution would have permitted squandering of resources on a scale necessary to build and maintain such a massively underutilized organ. Moreover, doubts are fueled by ample evidence from clinical neurology. Losing far less than 90 percent of the brain to accident or disease has catastrophic consequences.

Today, thanks to EEGs, PET scanners, MRI technology, magneto encephalographs, and other technologies that allow neuroscientists to observe what is happening beneath our skulls, we now know that we use *all* of our brain. *We just don't use it all at the same time.*

In the words of Dr. Elkhonon Goldberg, clinical professor of neurology at the New York University School of Medicine, "New neuro-imaging methods have changed neuroscience in the same way the telescope changed astronomy." For the first time in history, researchers can observe, in real time, which parts of the brain become more active when we solve problems, fall in love, tie our shoes, and run a marathon. This advance alone offers modern civilization an enormous advantage that ancient societies didn't have: *the potential to understand the human brain well enough to arm it against a recurring cognitive threshold.*

In fact, by studying the human brain at work under a variety of conditions, neuroscientists today have already uncovered three powerful antidotes to reaching a cognitive limit: brain fitness technology, the evolution of insight, and valuable new information about the unconscious mind.

Compensating for Cognition

On a foggy morning in 2007, the first gymnasium aimed at brain-training opened on Sacramento Street in San Francisco, California. vibrantBrains was the inspiration of Jan Zivic and Lisa Schoonerman, two entrepreneurs determined to make available to the general public what neuroscientists now understand about how to improve the brain's natural abilities.

Schoonerman sees brain fitness as a missing component of the overall health fitness movement:

> The demand for anything which might stave off dementia and Alzheimer's or memory loss resulting from old age is huge. People want to do something to protect their ability to think, then once they get here, they see that it's more than just protecting what you already have. The brain can build new circuitry anytime. It's really remarkable that you can learn new things, new ways of thinking no matter what your background or age.

Schoonerman has it right: Growing numbers of people are concerned about losing their cognitive abilities and want to know what they can do to maintain the health of their brains. As a result, vibrantBrains has become an overnight success. A second facility was opened in Foster City, California, and plans to to expand are under discussion.

But exactly how big is the trend?

In a recent *Wall Street Journal* article, reporter Kelly Greene interviewed Alvaro Fernandez, co-founder of SharpBrains, who revealed that Americans spent a stunning $80 million on mental fitness products in 2008. And despite tough economic times, the demand is climbing fast. As neuroscientists amass more evidence that the human brain can be manipulated in deliberate ways to safeguard against cognitive diseases such as dementia and Alzheimer's, brain fitness is finding its way into mainstream culture—in retirement communities, schools, hospitals, and households. Facilities similar to vibrantBrains have already sprouted up in Boca Raton, Florida (Sparks of Genius), and in California and Texas (Nifty after Fifty). New Web sites dedicated to cognitive fitness are quickly populating the Internet, not to mention a growing number of gaming products claiming to enhance a person's "brain age." As a large population of baby boomers face the realities of growing older, the desire to hold on to our ability to process information and to remain alert intensifies. *No matter how fit our bodies are, we now realize that when we lose our brains, life as we know it is over.*

To prove that brain fitness works the same way physical workouts do, vibrantBrains cleverly modeled their program after traditional health clubs. Clients join vibrantBrains the same way they become a member of a gym: Sixty dollars a month buys an annual membership that allows clients access to a facility where they use a series of computer-generated exercises for the brain called the Neurobics Circuit. The Neurobics Circuit consists of specially designed cognitive exercises designed to improve memory, visual and spatial focus, reasoning skills, alertness, and reaction speed as well as enhance concentration. The training regimen also includes exercises for stress reduction, mood elevation, and relaxation—skills also proven to enhance our problem-solving abilities.

Clients choose their own workout schedule and visit the friendly facilities as often as they like. It has become common for members to stop in before or after work, alternating days between a physical workout at their health club and an equally vigorous mental workout at vibrantBrains.

One of the unique features of vibrantBrains is an easy computer program that allows clients to keep track of their progress as their cognitive abilities, such as problem solving, improve over time. This feedback offers members the opportunity to "fine-tune" their individual workouts in order to strengthen areas where they have lower scores; similar to athletes who select equipment and exercises to strengthen particular muscles in the body, each software program is designed to challenge different brain functions.

For example, one of the computer programs offered by vibrantBrains begins with a simple Disney-like underwater scene that looks like the inside of an aquarium: There's an assortment of seaweed and tropical plant life, some coral and sand, and so forth. Suddenly a jewel appears, followed by three identical fish. One of the fish quickly conceals the jewel on the screen. Then all three of the fish begin swimming in random patterns—sometimes crossing over each other and sometimes swimming in opposite directions. Suddenly all the fish stop. The goal is to identify the fish that is concealing the jewel—a task that is relatively easy to accomplish when there is only one jewel and three fish.

Over time, however, more and more identical fish appear, and they begin swimming in increasingly erratic and complex patterns—and also faster. The task of keeping track of the jewel hidden behind one of the fish becomes more difficult. Then later more jewels also appear, so more than one fish is hiding more than one jewel.

It doesn't take long before our brains go into complete failure. And once we hit our cognitive limit, we immediately sense it. There are simply too many fish hiding too many jewels. We can't keep track of them all.

Yet, similar to our experience with video games, the more we practice, the better we get. Even though there are no patterns to memorize, surprisingly, our brain eventually learns to track three, four, and five fish hiding two, three, and four jewels. As the visual complexity of the game increases, everything from spatial focus, concentration, reaction time, and visual interpretation becomes enhanced. Our brains are getting a workout and getting more agile, quicker, and stronger by the minute. In this way, brain fitness works as a painless, fun way to develop "faster thinking, sharper focus and better memory," all skills that counteract the cognitive threshold.

The impetus behind vibrantBrains' Neurobics Circuit is "Jewel Diver," a brain fitness program developed by pioneer Dr. Michael Merzenich, neuroscientist and professor emeritus at the University of California, San Francisco. Merzenich is a spitfire who tirelessly travels around the world lecturing on the proven benefits that brain fitness is having on millions of schoolchildren and senior citizens. Armed with bulletproof clinical data and statistics, Merzenich is on a mission to change the way we teach and learn. In his view, we now understand a lot more about how to "prepare" the brain to accept new information more easily and permanently. So why not put what we understand about the human brain into practice?

But Merzenich's claim that brain fitness can be encouraged by exercises that escalate in sound, vocabulary, and visual complexity is the subject of considerable debate—primarily because he formed his own company, Posit Science, to produce and distribute software products like Jewel Diver. Many of Merzenich's colleagues worry that he has a "conflict of interest," and they also suggest that the results of brain fitness are exaggerated. Merzenich, however, takes a practical view:

> The tools we have may not be perfect, but we know they work to some degree. Just because we don't have data to determine exactly *how much* the technology helps doesn't mean we don't already know that it works. We've already proven this. So whether it helps a lot or a little, why not start using it? Over time, as we learn more, the products will keep improving.

Is there independent evidence that Merzenich is on the right track? Any data that suggest that we are on the cusp of *volitional cognition?* It turns out there is.

For example, in 2007, three major universities along with the Mayo Clinic participated in a study on the effectiveness of brain fitness. Science writer and editor of *Faith in Science* Gordy Slack reports on the results of their research:

> Half the group did the Brain Fitness program for at least an hour a day, five days a week; the other half watched educational DVDs. After 8 to10 weeks, the groups took a variety of learning and memory tests, including one in which they were asked to recall details from a story or words from a list. The test showed that *the memories of those in the Brain Fitness Group improved, on average, about a decade's worth* compared with those in the other group, who just did the watching.

A decade's worth of improvement? Quick, sign me up.

But that's just the tip of the iceberg. Additional studies performed at the W. M. Keck Foundation Center for Integrative Neuroscience at UCSF showed that participants who were over fifty years old and used brain fitness programs achieved the "neurological performance abilities" of thirty- to thirty-five-year-old adults. Reverse our brain age by twenty years? Is it any wonder neuroscientists like Merzenich question why brain fitness isn't front-page news? Why it isn't already part of a daily health regimen?

What's more, there's mounting evidence that brain fitness keeps paying dividends long after we give our brains a workout. The effects of brain fitness may not, as we once thought, be temporary. They may be permanent.

Nine years ago the National Institute of Aging (NIA) began testing a "Useful Field of View" (UFOV) developed by Karlene Ball and Dan Roenker. Similar to Merzenich's brain fitness programs, UFOV exercises were specially designed to increase reaction time to objects moving in a person's peripheral view without disturbing their "center of gaze." Three thousand volunteers were asked to use the UFOV system for a total of ten hours. The volunteers played a series of games where the complexity of tasks was gradually increased by adding more things to their peripheral vision. Follow-up studies proved that *five years later,* after spending only ten hours using the programs, the volunteers still retained benefits. They were able to process much more information from a wider field of vision than their counterparts. The volunteers also showed greater memory, reported fewer automobile accidents, sustained independent living for a longer period of time, and remained physically healthier than the control group, which had never been exposed to brain fitness.

The *Journal of the American Geriatrics Society* also confirmed the findings: A study of 487 adults age sixty-five and older who used brain fitness programs for forty hours over an eight-week period also revealed long-term improvements in memory, response time, concentration, and problem-solving.

In fact, there are now hundreds of clinical studies worldwide reporting similar results. Whether young or old, educated or not, brain fitness has a measurable, lasting effect on the cognitive skills of the human brain. Our brain becomes more facile, quicker, and better able to learn when we continuously challenge it—but not just by doing crossword puzzles and watching *Jeopardy.* We have to challenge it in very specific ways. And we are getting very close to understanding exactly what those ways are. Word by word, we are now constructing a user's manual to the human brain—ground zero for everything we do, know, think, and feel—and the sustainability of human progress.

Pouncing on Plasticity

On the international stage, Merzenich is best known for his pioneering work on a phenomenon called brain "plasticity." Plasticity simply refers to the brain's ability to burn new circuits whenever it chooses to. Merzenich was one of the first neuroscientists to prove that *at any age,* the human brain can decide to create *new* circuits to replace (or operate in parallel with) the old ones. This is really useful when the old ones stop working well, which it turns out begins occurring in humans around the age of *thirty years old.*

(Yes, that's right: We start losing cognitive capabilities around the time we turn thirty.)

The discovery of brain plasticity was an unprecedented breakthrough in neuroscience because it proved that the brain could be "rewired" following a debilitating injury. So when the brain cells once responsible for moving our right arm became damaged, our brains had the ability to program other cells to move that arm. Brand new circuits could be developed to perform new tasks.

Merzenich provides an easy example of how the human brain learns new tricks:

If you are a young boy growing up in Sao Paulo, Brazil, there is better than a 50 percent chance you can run and bounce a ball on the top of your head at the same time. At an early age, the brains of Brazilian boys learn to perform this feat with great ease—*because this is what boys in Sao Paulo do.* But in downtown Minneapolis, you don't expect to see this behavior. In fact, you'd find very few boys capable of running while bouncing balls on the tops of their heads. Why? Because their "social circumstances" did not elevate this skill to become a suitable *goal* for the brain. No perceived advantage means no circuits being constructed by the brain to *run and bounce.* Can a boy in Minneapolis *learn* to run and bounce a ball on his head? Of course he can. In fact, one of the amazing attributes of brain plasticity is that we can choose to learn new skills throughout our entire lives. Unlike other animals, the human brain remains *plastic* until the day we die.

Then why don't boys in Minneapolis develop the same cognitive-muscular abilities as boys in Sao Paulo do? Simply put, the brain starts *deciding* how it wants to evolve by observing the social environment and selecting goals it thinks are good for it and for its group. Once the brain picks a goal, it also sets itself up to judge whether that goal is being successfully achieved. Every time a boy in San Paulo makes a new attempt to bounce the ball on his head, the brain whispers, "Good job! Save that one in memory!" or "Nice try, but don't do it that way again." Similarly, if boys in Minnesota *decided* these skills were important, their brains would begin burning circuits for running and bouncing in this same way. This is how our brains learn.

Dr. Michael Stryker, a colleague of Merzenich's at the W. M. Keck Foundation Center for Integrative Neuroscience, elaborates on how the brain prioritizes what it wants to learn: "Parts of your brain are engaged in a constant competition to see which ones [neurons] can create lasting changes. And where you focus your attention helps determine where the positive plasticity takes hold."

It all boils down to this: Neurons that fire simultaneously in the brain form circuits that the brain later draws upon to perform tasks and solve problems—or, as author Gordy Slack once put it, "Neurons that fire together wire together." So, when faced with mounting complexity, the real challenge is to unlock ways to form new circuits—new ways of thinking—as often and as quickly as possible.

Here Merzenich is quick to point out that the first rule governing problem-solving is that *we can only draw from what our brains have inherited, experienced, or learned.* This is the raw material for all of our thoughts, ideas, innovations, and solutions—everything we know and imagine.

For example, if we have never learned anything about mathematics, physics, or engineering, the next big breakthrough in particle physics is not likely to suddenly "drop out of thin air" and surprise us. We aren't going to suddenly wake up one morning and start scribbling down the equivalent of Einstein's theory of relativity. That's because our brains simply don't have the *content* to do so. It doesn't matter how intelligent or creative we are. The type and amount of content we load into our brains has a large effect on the solutions we can develop to complex problems.

In other words, the human brain doesn't create solutions out of nothing—not even creative ones. Our brains have to have raw material to work with, even if the raw material can't be detected in the final solution. For instance, it's unclear which specific content in Wag Dodge's brain caused him to have a life-saving insight in the middle of the Mann Gulch fire, but it is clear that he must have had resources to draw upon that fifteen other firefighters who lost their lives did not.

In truth, depending on how much information our brains process, absorb, and inventory, we sentence ourselves to shopping in either a megawarehouse stacked to the ceiling with options or a small boutique with a limited selection. This is the reason a successful education is so important during the early years of childhood: *The better learners we become, the bigger the store our brain can shop in for the rest of our lives.* Or in the words of Louis Pasteur, "Chance favors the *prepared* mind." The more prepared we are to have an insight, the more likely an insight will occur. So, learning to *efficiently load content is a key component of improving our odds of having an insight.*

This is why Merzenich's breakthrough in brain plasticity is historic. When learning becomes the habitual state of our brains—a naturally occurring part of everyday life—forklifts keep loading the brain with new "circuits" until the day we die. This is one of the truly unique and exciting features of the human brain: It's never too late to turn a small convenience store into a warehouse jam-packed with content. All we have to do is take a class, read a book, go for a walk, learn a new dance, or adopt brain fitness to begin burning new circuits.

New Tools for Schools

When children "learn to learn" at an early age by using brain fitness, the results are astounding.

Merzenich noted that, today, over 1.5 million elementary and high school children in the United States are already using brain fitness technology to improve language, reading, and overall cognitive development. The new "Fast Forward" software program, designed to "improve the accuracy and speed at which a child can accumulate information," is getting results far beyond what increasing teacher's salaries, buying new textbooks, or administering tests to students have achieved. Imagine that. Playing a simple video game a couple times a day may turn out to be all the boost schoolchildren need to learn at a faster pace.

Merzenich has discovered that children who use brain fitness programs for a short period of time prior to the start of classes have a substantially easier time with processing and absorbing new content. And the more content children retain, the more resources they have to draw on—no matter whether they are solving problems using left- and right-brain methods or insight. He sees remarkable similarities between brain fitness and warm-up exercises that athletes regularly perform before competition. Like the body, the human brain wants to be *warmed up* before it exerts itself. And the way to warm up the brain before exerting it is to ease it into a pattern of learning.

As an example, in Jacksonville, Florida, approximately twenty-three thousand children who were exposed to a total of fifteen to seventeen hours of brain fitness have now been studied for over three years. The results are startling. Children who were exposed to brain fitness showed *twice* the academic achievement as children who were not exposed. What's more, the cognitive advantages appear to grow exponentially each year. It's as if brain fitness offers advantages that continue to multiply over the entire arc of a human being's life. Once you get the brain started in the right direction, it just keeps going and going. Talk about leverage.

Injecting Insight

But does brain fitness go far enough?

It may be an important component of catching cognition up to complexity, but is improving our ability to absorb more content useful if we can't use it to our benefit? After all, the definition of intelligence is twofold: the ability to acquire knowledge and *the ability to apply that knowledge to problem-solving.*

So, more important than *how much* we learn may be *what we do* with what we learn. To this end, a curriculum that continues to reinforce left-brain analysis and right-brain synthesis problem-solving may only succeed in propagating more of the same. And as we discovered earlier, it doesn't take long before both of these methods become overwhelmed by the magnitude of complexity we must contend with.

That's where *insight* steps in.

According to Drs. Kalina Christoff, Alan Gordon, and Rachelle Smith, authors of a recent study titled "The role of spontaneous thought in human cognition" (University of British Columbia), the answer lies in "spontaneous thought": "Spontaneous thought facilitates the process of making sense of our experiences, the drawing of connections between memories and concepts, the broadening of attentional focus to include larger amount[s] of information into consideration, and the process of assigning motivation value to experiences—all factors that are essential in making a good decision in a complex situation."

They continue: "The process of deliberately thinking things out may be a part of the decision process, but it is the more spontaneous, defocused thinking mode . . . that may be necessary for important decision[s] to be successfully made."

In short, *insight*—the next important step in human cognition—is antithetical to the methodical, analytical thinking we have relied on for centuries.

Merzenich has the right idea about the importance of making learning the *steady state* of the human brain. But it is equally important to avoid highly structured, uniform curricula that treat every student the same way. To engender *insightful* problem-solving, we must begin rewarding students for spontaneous, correct answers. To prepare students to successfully navigate an increasingly complex world, we must teach them to make novel connections rather than continue to rely on reductionist thinking.

According to MIT PhD and former editor of the Public Library of Science (PLoS), Dr. Hemai Parthasarathy, people who have insights report a sudden, unexpected ability to "see connections that have previously eluded them." So the real opportunity in education is not only to increase a student's ability to load content but also to stimulate spontaneous, insightful problem-solving.

This is where the work of neuroscientists who are studying insight plays a critical role in the future of humanity: Insight holds the key to breaking a cognitive threshold that has been responsible for the collapse of every advanced civilization. It is a proven method for unraveling otherwise unscalable levels of complexity.

A New Way for a New Day

There's no question that insight has been around for a while, but it's only in the past decade that we have been able to observe how it works in the human brain. In fact, it's so new that we have many more questions than we have answers about when and why it occurs.

That said, what we have uncovered to this point gives us great hope that insight can bridge the cognitive gap between the physical evolution of the brain and growing complexity.

The first thing we now know is that insight exists.

It has not only been observed and documented; it has also been photographed working in the human brain. The brain's activity during insight looks considerably different from normal left- and right-brain problem-solving. We also *feel* differently when we call on insight. According to Dr. Karuna Subramaniam of the University of California, San Francisco, "Analytic processing involves deliberate application of strategies and operations to gradually approach a solution. Insight, which is considered a type of creative cognition, is the process through which people suddenly and unexpectedly achieve a solution through processes that are not consciously reportable."

He continues, "Insight solutions tend to involve conceptual reorganization, often occurring after solvers overcome an impasse in their solving effort, and are suddenly able to recognize distant or atypical relations between problem elements that had previously eluded them. . . . When [a] solution is achieved, these factors combine to create a unique phenomenological experience, termed the Aha! or Eureka! moment."

Second, insight is a biological phenomenon.

It's not "weird science" or some mystical experience but rather a naturally occurring physical function in *every* human brain. Regardless of our experiences, education, or background, people of all ages, from all walks of life have insights. It's nature, not nurture.

Third, unusual areas of the brain become activated when insight is used to solve complex problems.

In the 2006 article "The Prepared Mind," Drs. Kounios, Jung-Beeman, and five other scientists from three leading universities describe a landmark experiment in which nineteen subjects were asked to solve word problems as the activity in their brains were monitored. In this experiment, three simple words, such as "pine, crab and sauce," were presented to each participant. The participants were asked to think of one word that could be combined with each of these three words to create three completely *new* words. For example, by adding the word "apple" to the words "pine," "crab," and "sauce," the new words pineapple, crabapple, and applesauce could be created.

Following each successfully solved problem, the test subjects were asked whether they experienced any of the spontaneous, epiphany-like sensations associated with *insightful* problem-solving.

It turns out that not only were the participants able to identify when they used insight instead of left- and right-brain problem-solving, the areas of their brain that became activated when they claimed they used insight were also different.

When a person used insight to solve a problem, an otherwise unremarkable fold in the brain called the anterior Superior Temporal Gyrus (aSTG) became highly excited: A sudden burst of gamma oscillatory activity was consistently recorded. The aSTG is a "crease" in the temporal lobe that, to this point, has been associated with processing sound and language and forming abstract associations. This little-known fold "lights up like a Christmas tree" when insight is used to tackle a problem. But today, all we have are hypotheses to explain why.

The same neuroscientists also discovered that the Anterior Cingulate Cortex (ACC), responsible for relaying signals between the left and right hemispheres of the brain, "appear[s] to suppress irrelevant thoughts" just prior to evoking insight. There was an effort by the brain to momentarily turn other worries and thoughts off in order to clear the way for extreme concentration. According to the study the "ACC may be involved in suppressing irrelevant thoughts . . . thus allowing solvers to attack the next problem with a *clean slate*. This explanation assumes that insight processing is more susceptible to internal interference than is noninsight processing, thereby necessitating greater suppression of extraneous thoughts."

This may explain why exercises such as meditation, yoga, and relaxation often lead to insights. The fact that Newton was sitting under a tree and Archimedes was lying in a bathtub when they had their insights may have been no coincidence at all. The mind may need to "wander" in order to make connections that "previously eluded" it.

Fourth, insight is extremely cognitively taxing.

In addition to the aSTG and ACC, the electrical activity in *four* other areas of the brain also significantly increased when insight was used. These increases were recorded for all of the subjects when they reported having the sensations associated with insight. With so much electrical activity in the brain before and during insight, neuroscientists hypothesize that insightful thinking may be too cognitively demanding for human beings to use for every problem. Insight may occur infrequently because it requires considerably more horsepower than logical analysis and synthesis. Perhaps, at this time in human evolution, insight simply isn't sustainable for long durations of time.

Fifth, the mind "prepares" itself to use insight in advance, and there appear to be specific conditions that facilitate it.

In "The Prepared Mind," Dr. Kounios noted that 300 milliseconds before insight was deployed, the brain could be observed preparing itself. Jonah Lehrer aptly summarizes Kounios' findings by saying: "A small fold of tissue on the right hemisphere, the anterior superior temporal gyrus, became unusually active in the second *before* the insight. The activation was sudden and intense, a surge of electricity leading to a rush of blood."

Not only have Kounios and Jung-Beeman traced insight to its source, it also appears they are able to predict when insight will be used to solve a problem *before* the problem is solved. "We can demonstrate the preparation for problem solving can be associated with distinct brain states, one biasing toward a solution with insight, the other biasing toward a solution without insight."

In other words, what we do before we attack a problem may determine whether we will use insight or not. This means that there are conditions that encourage the use of insight as well as conditions that prevent it.

If this is true, what did Wag Dodge do to prepare his mind to have a life-saving insight? What were the conditions that led to breakthroughs by Einstein, Charles Townes, Richard Feynman, or James Watson? And if there are conditions that facilitate insight, does this mean we can deliberately create those conditions?

Unfortunately, we just don't know. At least not yet.

Finally, *insight is the human brain's special weapon against complexity.*

When it comes to highly complex problems, insight succeeds where left- and right-brain problem-solving methods fail. Given what we now know about the process of natural selection and how organisms strive to adapt to changes in their environment, we can assume that an increasingly complex environment will have an effect on the evolution of the brain. If we know that insight is ideal for managing complexity, then it also follows that this capacity will continue to grow rapidly over millions of years.

Whereas human evolution to this point has been driven primarily by environmental change, for the most part we now control the physical environment in which we live. When we don't like our natural environment, we simply change it. We chop down a forest. We install dams and bridges and extend the shorelines with landfill. We tunnel through oceans, defy gravity in outer space, and produce clothing and shelter to endure sub-Arctic temperatures. We climb the highest peaks and descend into the skull-crushing depths of the ocean while carrying the oxygen we require on our own backs. We conquer all these hostile environments with no need for biological change: no new appendages, no nothing.

So, barring radical adjustments forced on us by extreme changes in the environment such as an ice age, human beings are now shaped by a changing *social* environment. In other words, the human brain is now adapting in response to a quickly shifting man-made environment, not a physical one.

In 2004, the University of Chicago Howard Hughes Medical Institute published an article describing the research of Dr. Bruce Lahn, professor of Human Genetics: "Why the human lineage experienced such intensified selection for better brains but not other species is an open question. Lahn believes that answers to this important question will come not just from the biological sciences but from the social sciences as well. It is perhaps the complex social structures and cultural behaviors unique in human ancestors that fueled the rapid evolution of the brain."

The authors of Serendip, a Web site that addresses the relationship between the brain and evolution, agree. The more complex our behavior becomes, the more impact that complexity has on our physical evolution: "*Behavioral Complexity* may relate to the extent to which the brain functions as an *amplifier of evolution.*"

In this respect, *insight can be viewed as nature's gift:* a brilliantly efficient way to cut through thousands of variables, multiple wrong solutions, and produce a correct and elegant answer. Whereas traditional left- and right-brain problem-solving methods become overwhelmed by complexity, insight soars right through chaos—much like a hyperefficient editor who instantly isolates knowledge that is essential to a solution from facts that are not.

This is essentially what happened to Wag Dodge and Newton and Einstein and Archimedes and Galileo and Townes and Chu and

The Characteristics of Insight

1. The solution arrives suddenly. It often occurs following a period of gridlock or feeling stuck.
2. The problem-solver has difficulty tracing the thought process leading up to the answer.
3. There is an overwhelming feeling that the answer is correct.
4. The answer *is* correct.
5. There is a sudden burst of gamma-band oscillatory activity in the anterior Superior Temporal Gyrus (aSTG).
6. Insight is a demanding cognitive process. Increased electrical activity occurs in four areas of the brain: left posterior M/STG, anterior cingulate, right posterior M/STG, and left amygdala.
7. Both the left and right hemispheres of the brain become activated.
8. The brain prepares itself 300 milliseconds in advance: The aSTG becomes excited, and the ACC begins shutting down distracting internal and external thoughts.
9. Insight is extremely efficient at solving highly complex and difficult problems, allowing us to see connections that have been previously overlooked.
10. Insightful solutions are unconventional, highly innovative, and far-reaching in their ramifications because they are free of supermemes.

Kamen. Everything they experienced and knew was quickly inventoried and then ruthlessly edited down to only those factors relevant to the solution—then *bam!* They connected the dots in a new way. *They had an insight.*

We may not understand exactly how insight works, and we may not know when or whether it will be used to solve a problem. But we do, however, know that in terms of understanding how the brain works when we are having an insight, many of the road signs point in the same direction—*toward the unconscious mind.*

Complexity and Consciousness

The connection between insight and the unconscious mind occurred to me one day when I was interviewing entrepreneur Jerry Lauch. Lauch is one of the early pioneers responsible for developing commercial sleep disturbance clinics that diagnose and correct abnormal sleep patterns. Because of his vast background studying the effect that sleep deprivation has on learning, decision-making, and human behavior, I often find myself picking Lauch's brain.

One day when we were discussing the difficulty neuroscientists are having identifying exactly how insight works, Jerry said,

> The important thing we are doing when we sleep is organizing and making sense out of information we already have or recently experienced.
>
> Our unconscious mind is a lot more powerful than our conscious one. When you think about it, that's how human beings manage high levels of complexity deep in the recesses of our unconscious. No wonder when we have an insight it feels sudden. No wonder we can't explain exactly where our epiphanies came from. No wonder we can't decide to have an insight whenever we want to.

So is insight a conscious or an unconscious process? Is the difference between left- and right-brain thinking and insight really the difference between conscious and unconscious problem-solving? After all, we know the unconscious mind is at work even when we are awake—just ask any psychologist. So, if insight is an unconscious form of problem-solving, how will our understanding of the unconscious mind contribute to our greater understanding of insight?

The unconscious mind itself has long remained a bastion of unscalable human complexity. Its exploration began with psychologists

William James and Boris Sidis, a student of James, who in 1898 published *The Psychology of Suggestion: A Research into the Subconscious Nature of Man and Society*. Later, in 1899, Sigmund Freud published his groundbreaking research, *The Interpretation of Dreams*, which launched the modern psychoanalysis movement.

Much later, thanks to E. O. Wilson's revelations in sociobiology and James Watson's discoveries in molecular biology, we learned that many of the human drives attributed to the unconscious mind were really inherited predispositions and biological instincts. We are not born blank slates: Evolution equips every human with motivations, skills, and strategies for survival. Even so, there is still an awful lot about the unconscious mind that can't be explained by evolutionary biology or Freudian psychology alone.

Despite this, we would be hard-pressed to find a scientist who does not acknowledge the fact that complexity favors the unconscious mind. The fact that insights arrive spontaneously and cannot be traced to any identifiable conscious process may indicate that insight is connected in some way to our deep unconsciousness. In time, as we learn more about the unconscious mind and the human brain, it's likely we will also achieve a better understanding about the inner workings of insight.

Insights Large and Small

Whether they stem from the conscious or unconscious mind, some insights are about small, targeted problems, such as solving the world's potable water problem, while others represent major paradigm shifts, such as the discovery of the double helix in DNA. Regardless of the size of the problem, all insights have a streamlined, immaculate quality to them. Like the insight that propelled Wag Dodge to burn a fire shelter around himself or the observations that led to Darwin's theory of Natural Selection, the problems may be complex, but the insights themselves are often elegant and obvious. Insights have a way of cutting to the heart of an issue by ruthlessly dismissing extraneous,

unessential information. In addition to being spontaneous, free of supermemes, correct, untraceable, and engaging the STG and ACC,
they are also characterized by a certain precision. Take, for example,
$E = MC^2$. It doesn't get any more succinct than that.

Not long ago, in what could have turned out to be a tragedy, I had
an opportunity to witness firsthand the repercussions of a simple, elegant insight—not from a Nobel award–winning scientist or famous inventor but from a neighbor and friend.

In the spring of 2008, a dry lightning storm sparked a fire in Big
Sur that quickly became one of the largest, most destructive forest
fires in California history. As coastal winds whipped the flames fourteen stories high, black smoke spread across the horizon like spilled
ink and a blizzard of hot burning embers began floating through the
air, landing on dry grasses and leaves. The uncontrollable blaze
jumped one fire line after another, and there appeared to be no way to
stop the airborne assault. The entire town of Big Sur was ordered to
evacuate. People grabbed photo albums, pets, quilts, and paintings
and fled north to the nearby town of Carmel-by-the-Sea, where residents welcomed them with open arms, setting up extra beds in their
guest rooms, offering motor homes in their driveways and vacant hotel
rooms free of charge. The local office of the American Red Cross immediately set up shelters at two schools, but not one evacuee checked
in. Friends, relatives, and helpful citizens in town absorbed them all.

As the fire consumed one home after another in an unstoppable
wave, it suddenly occurred to me that the home of my friend John
Saar was in the fire line.

John is a local fixture. He's lived on the Big Sur coast for most of
his life and has owned a successful business in town for as long as anyone can remember. He's the kind of fellow who can't go anywhere
without stopping to talk to five people and pet three dogs, all of whom
he knows by name.

As soon as I realized the fire was headed his way, I called John to
see how he was doing. But there was no answer. So I kept calling—
wondering if maybe he could use an extra truck or pair of hands to
move things out of his house. A day later, still no return call. And two
days later, the same thing.

Then on the third day, when it looked like the fire was either closing in on his home or had already consumed it, John came driving up
my driveway. I rushed out to greet him.

"Where've you been? What happened to the house?"

"The house is fine," he said as he dusted ashes off his pants.

"Well, what happened—I've been trying to reach you for days."

"Well, I'll tell you. I was up there, you know, at the house—watching the fire and smoke from my deck, and, well, I was pretty sure it
was coming toward me. I guess I was getting myself used to the idea I
was going to lose the house. I was just sitting there, you know, just en-
joying it for one last time. I already had the car packed. But I thought
I might as well sit there as long as I could—you know, have a beer and
watch the flames cross the hill."

"Yeah?"

"So I kept thinking about what I could do, you know, to try to save
the house. I already knew the firefighters weren't going to be able to
help me. They're trying to keep the town from burning up and the
wind was getting bad. So I just kept sitting there looking at all the
smoke and then for some darn reason I started thinking about 'firewalls.'"

"Firewalls?"

"Yeah, you know, the kind they make you build between your
garage and your house. It's part of the building code now. You have to
build a firewall and put in a fire-rated door between your garage and
house."

"I didn't know that."

"Then, you know what? I thought, 'Hey, a firewall is basically just a
double layer of sheetrock.'"

"Really?"

"Yeah. That's all it is. So, I had some pieces of sheetrock left over
in my garage from some remodeling I did last year, and I started
breaking off pieces and throwing them in my fireplace. I tried to burn
them. I left them there for a little bit and then I took them out. You
know what? That stuff doesn't burn. You can't make sheetrock catch
fire. Not even if you throw it *in* a fire and leave it in there."

"So?"

"Well, I'll tell you. As soon as I figured that out I grabbed some guys and we ran down to Home Depot and Ace Hardware with three trucks and we bought all the sheetrock we could carry. I hired a bunch of laborers and drove them to my place and we started covering the outside of my house with sheetrock as fast as we could."

"What?"

"Yeah, the roof, the decks, all the walls, the windows, the whole thing. The place looked like some weird space-aged bunker. People were driving by and taking pictures of it."

"You covered the whole house?"

"Yeah. The whole thing. Took about three hours. Then, when the embers started blowing over the hill, I watched them land on the sheetrock. They just burned out. It was unbelievable. Nothing happened. The house is fine, except it's weird to go inside it. It's dark like a cave."

Two things were notable about John's spontaneous revelation: His unplanned insight saved his home—even while other homes around him burned—and second, remarkably, John's insight was free of supermemes: no irrational opposition, personalization of blame, no false correlations, silo thinking, or economic considerations. Though he probably could have, John didn't once try to leverage his discovery into some profitable enterprise. Nor did he weigh the benefits of nailing sheetrock to the outside of his house against the cost of an unproven last-ditch effort.

The fact is that John set aside every modern supermeme. Similar to Mohammad Yunus and Dean Kamen, he was free to think outside the box, to let his mind wander until it bumped into a solution parked somewhere in the recesses of his mind; then suddenly he discovered a novel connection between firewalls, sheetrock, and protection of the exterior of a house from an encroaching fire. And at that moment in time John's solution was no less important to him than the one that once saved Wag Dodge's life.

Real change—change that matters—rarely comes from business as usual.

Today, there is evidence that even under the best of circumstances, only about 4 percent of the population, regardless of education or background, taps into the advanced problem-solving capabilities of the brain. According to science writer Jonah Lehrer, in simple tests where people are given a box filled with a piece of cork board, a candle, and matches and "asked to attach the candle to the cork board so it will burn properly," almost 90 percent of the people choose two *incorrect* solutions. Only around 4 percent realize the cardboard box is itself the key to solving the problem.

How many of those 4 percent capable of managing complexity hold important government posts or are at the head of global conglomerates today?

William Futrell, a pioneer in educational reform who is working diligently to bring new learning techniques to impoverished inner-city schools, recently summed up his frustration with how long it takes solutions—even proven ones—to be acknowledged and to take hold:

> I keep thinking that it was 700 years after the invention of the saddle before someone thought of the stirrup. With the invention of the stirrup man could change his weight and momentum to force a horse to charge and suddenly, the power structure of whole civilizations changed. Amazingly, it took another 200 years before someone thought of building a wall to stop the charging horses. Ideas come slowly, and often a simple idea changes everything. But where are they?

Sometimes important insights come from throwing pieces of sheetrock into a fireplace. Sometimes they come from an apple that drops from a tree, and sometimes they come from water spilling over the edges of the bathtub. But no matter who has the insight or under what conditions, insights have widespread ramifications.

Taking the Next Step

Looking backward, we know that the human organism has succeeded in making four critical leaps in evolution and that we may now be on the cusp of a fifth.

Insight is nature's answer to the cognitive threshold—the bridge between the slow evolution of the brain and problems that have become too complex for us to understand or solve—a recurring paradox that has held civilizations hostage for centuries.

As the human brain makes an all-important adaptation to an increasingly complex, challenging cognitive environment, insight will continue to evolve. Over millions of years, the methods of problem-solving suited to high degrees of complexity will become perfected—first by occurring infrequently and spontaneously and then much later by becoming such a part of everyday life that insight will be indistinguishable from left- and right-brain methods.

The real question is: While Mother Nature is racing to meet our growing need, can we leverage what we now know about insight to give evolution a helping hand?

— 12 —

Invoking Insight

Conditions Conducive to Cognition

Just last month I was invited to speak to a group of students about the relationship between complexity, collapse, and correction.

A few of the students were interested in the rise and fall of ancient civilizations, but many more of them wanted to know whether I saw any signs of a cataclysmic collapse around the corner.

"Sorry, no such luck," I told them. "You have plenty of time to finish your classes and graduate."

Although the threat of collapse garners a lot of attention, it's really just another opportunity to take a snapshot of how the human organism is doing at this moment in time: Where do we think we are in terms of evolution? And where do we stand in the historic pattern of decline? The beginning? The end?

After each presentation I encourage the audience to answer these questions for themselves in a spontaneous, no-holds-barred discussion.

Most of my exchanges with students are friendly and relaxed. We poke fun at and bemoan the foibles of being human, especially in a serious world bombarded by technology, information, and danger. We share an understanding that we are biological creatures who get out of bed every day and try to do our very best. Some days we do a great job, and on other days, not so much. But with less than a 2 percent difference in genome from our nearest relative, the bonobo chimpanzee, we have to admit that we have a lot more ambition than we have the biological apparatus to support. This fact alone seems to take a lot of pressure off of everyone in the room.

On this particular day, following the lecture, a large fellow stood up in the back of the theater and shouted in an exasperated voice, "Aren't you going to tell us what to do?" and everyone broke out in laughter.

Do?

I thought I had already explained what we could do: Acknowledge the pattern of collapse, guard against supermemes, buy time using venture capital models to mitigate, and embrace brain fitness to buttress our lagging cognitive abilities. In tandem with these things, we can vigorously pursue insight—nature's evolving antidote to the cognitive threshold. Neuroscience holds the key to modern man's survival.

I thought the cure was clear.

But once the laughter subsided, the student in the back of the theater persisted. What he really wanted were practical measures he could act on when he walked out the door. "If the key is to have more insights," he asked, "don't you need to tell us how to do that?"

He had a point. What good was all the theoretical mumbo jumbo if it didn't amount to something useful, something actionable?

Despite my reluctance to compile a "how-to" list for encouraging insight, I confess there are certain conditions that neuroscientists agree facilitate cognition, and some of these conditions, it turns out, also favor insight. But that's no big stretch: When we increase our opportunities to forge free connections in the brain, it makes sense that we are creating an environment conducive to better problem-solving. This doesn't guarantee that we will have an insight, but early evidence suggests we may better our odds.

However, before I launched into a list, I offered the students one important reminder: *Insight is a recent discovery.*

With little research on the subject, the data supporting the conditions favorable to insight are not only thin, they haven't been corroborated yet. Therefore, at the risk of contributing to more "counterfeit correlation," I can only offer what we *believe* today—subject, of course, to be proven or disproven by the *facts* of tomorrow.

Wise About Size

The human race has had a long and romantic tradition of crediting great breakthroughs to the tribulations of single individuals: Einstein, Archimedes, Benjamin Franklin, Van Gogh, da Vinci—the list is long. Interestingly, every story about an epiphany is similar: An eclectic individual attempts to solve a complex problem that has stumped humankind for centuries, when suddenly he stumbles on an "aha!" moment. At first his revelation is rebuked by everyone, and then, over time, the inventor is redeemed and his ideas enthusiastically embraced by all. The town throws a parade. The discoverer is praised for his unwavering persistence, lavished with acknowledgment and riches, and then rides off into the sunset as the film fades to black.

Nice story.

But is it true? Are insights the unique providence of individual genius?

Sometimes they are, but more often they are not.

It turns out there's a reason schoolchildren can name the inventor of the cotton gin and steam engine (even though they have never seen one) but have no idea who invented the Internet, semiconductor, or cell phone despite using these products every day. Today more and more innovations are the result of collaborative teams, not individuals. In fact, so many people play a role in complex collaboration that we would be hard-pressed to decide which individual's contribution was the most valuable: Was the mathematician more important than the chemist or physicist or engineer? It's as difficult as deciding who is responsible for a successful movie: The director? Producer? Screenwriter? The actors, the animators, or the editors? No wonder acceptance speeches for movie and science awards sound an awful lot like roll calls as recipients thank dozens upon dozens of collaborators and invite ten people to stand with them on the stage. They know that complex projects require collaboration—no single mastermind can claim all the credit.

That's because the solutions to large, complex problems demand a convergence of many different areas of expertise. No one person possesses enough talent in enough disciplines to tackle today's systemic problems. One person solve global warming? Not a chance. Terrorism? No way. Preventable childhood diseases? Not even Bill Gates himself. It's not a matter of resources; it's a matter of difficulty.

In fact, as complexity grows, the time required for a single individual to make a discovery is growing longer and longer.

In a disturbing article, science writer Robert Roy Britt reports that even the most dedicated and gifted among us are getting bogged down: "A study of Nobel Prize winners in 2005 found that the accumulation of knowledge over time has forced great minds to toil much longer before they can make breakthroughs. The age at which thinkers produce significant innovations increased about six years during the 20th century."

While it's true that a deluge of new information and technology has tacked years onto individual discovery, that doesn't necessarily mean major breakthroughs take longer to achieve. It may just mean that they take longer to achieve when we work alone.

Recent research reveals that the solutions to difficult problems are superior when people work in small groups rather than working individually. In 2006, the American Psychological Association (APA) reported in the *Journal of Personality and Social Psychology* that groups of three, four, and five people solve highly complex problems better and faster than individuals. A study of 760 students at the University of Illinois at Urbana-Champaign compared groups of various sizes to determine how long it would take for each group to produce correct answers to a problem. This and other subsequent studies revealed that groups of two or less do not appear to have enough critical mass and groups over five often become too cumbersome to solve complex problems efficiently.

According to writer Antony Jay in his book *The Corporate Man*, historically the ideal group size has been around ten individuals. Jay makes the case that small groups are efficient because for millions of years humans successfully worked together as hunters in groups of *about ten* persons. He further points to the success of the Roman and British armies that organized into small units of approximately ten soldiers each.

In 1998, Dr. J. Dan Rothwell, professor of communications at Cabrillo College, also demonstrated that the decisions made by small groups were superior to those made by individuals. Rothwell discovered that the minimum group size required to efficiently solve tough problems was three persons, but the maximum size was subject to the type and difficulty of the task.

Later, research by Drs. Tata and Anthony revealed that increasing group size produces more discussion and exploration, but also has one limiting side effect: The more people added to a group, the more opposing opinions and ideas were suppressed. When it comes to tolerating dissent, it turns out small groups are better than larger ones.

The disadvantages of large groups were also confirmed a year later by Drs. Harris and Sherbloom when they discovered that large groups often led to "social loafing" and "tagging along" by individuals who made no contributions to group decision-making.

So, what's the optimal group size for solving complex problems? From all accounts it's *greater than three and less than ten.*

It turns out that increasing our odds of innovative problem-solving might be as simple as working in smaller "neural groups" of four to nine people. Imagine the impact on the quality of our decisions if juries were reduced to teams of four to nine, or if corporate and non-profit boards were also trimmed. How about the size of government subcommittees, the number of students in a classroom, or the number of officials allowed at a table during peace talks?

But perhaps nowhere has group size played a greater role in determining success than in modern business. Group size may provide the single most important clue as to why start-up companies out-innovate and out-maneuver corporate giants in spite of the fact that these giants have significantly more resources at their disposal with which to crush their puny rivals.

Larger corporations, however, are encumbered by their own bureaucracies, protocols, politics, procedures, personnel, silos, and marching orders, often making them as inept as a muscle man furiously swatting at a fly. By contrast, nimbler start-ups engender greater creativity, faster development, and leaner and meaner operations because people work every moment of every day in small groups. With shallow pockets, inferior equipment, and fewer experts among them, small start-ups consistently achieve what no large conglomerate can, proving time and time again that group size trumps resources.

For this reason, large corporations have recently begun to view the acquisition of start-ups as an essential strategy for staying ahead. Although large companies may not be able to successfully organize into smaller groups, they have the wherewithal to purchase innovation from smaller companies, as demonstrated by medical giant Johnson & Johnson.

Between 1995 and 2004, Johnson & Johnson acquired 51 smaller companies. At the present time, J&J operates more than 200 separate subsidiaries—not divisions or departments, but rather independently operated smaller firms. Furthermore, their appetite for acquisitions continues to grow. Why? A quick look at J&J's top-selling drugs tells the story. Of the top six revenue-producing pharmaceuticals, only two were developed in-house by the pharmaceutical giant. The remaining four were acquired from smaller firms.

Dot-com legend Google is on the same path. Despite having a reputation for being one of the most innovative companies in the world, between 2001 and 2009 Google acquired over fifty-three smaller companies (about seven a year) to maintain their technological edge. Similar to J&J, the vast majority of Google's revenues are the result of meeting customer needs through acquisitions. Even an innovator like Google has been unable to leverage its vast resources, personnel, technology, and partnerships to outperform its nimble competitors.

If it's been proven that groups between three and ten people innovate and navigate complexity better than larger groups, what impact would working in "neural groups" of four to nine have on other types of organizations? How would it affect future global summits, rescue teams, and major league sports? And how about classrooms?

Interestingly, over twenty years ago the renowned Tennessee Student/Teacher Achievement Ratio (STAR) program revealed that "smaller kindergarten classes of 13–17 students were approximately one month ahead of their counterparts in classes of 22–25 at the end of one year, and by the second grade, those [students] in smaller classes were about two months ahead."

Once again, smaller groups also appear better for learning. So why not try reducing class sizes from seventeen students to four to nine children—the optimal size according to research. A case could easily be made that if we did nothing but introduce brain fitness programs into the curriculum for a few minutes each day and reduced class sizes, many of our problems in public education would subside.

In the final analysis, when it comes to insightful problem-solving, we need only think like Goldilocks: Some groups are "just too big," and others are "just too small." And as far as the human brain is concerned, *groups* of four to nine turn out to be "just right."

Cobblestones and Cognition

Have you ever wondered why one of the most difficult things to teach a robot to do is to *walk* on two legs?

It turns out there's a reason. Apparently, the simple act of walking turns out not to be so simple after all.

Professor Florentin Wörgötter of the University of Gottingen in Germany explains why teaching a robot to walk on bumpy terrains like cobblestones is so challenging: "Releasing the spring-like movement at the right moment in time—calculated in milliseconds—and to get the dampening right so that the robot does not fall forward and crash. These parameters are very difficult to handle."

Wörgötter elaborates further on the challenge a simple change in surfaces presents: "When it comes to more difficult things—such as a change of terrain—that's when the brain steps in and says "now we are moving from the ice to sand and I have to change something."

Whereas the human brain interprets surface changes and adjusts the body in rapid fire, it is extremely difficult for a robot to make these same lightning-speed calculations without toppling over. That's because walking upright on irregular surfaces is a cognitively intense task. Our eyes have to visually assess the height and depth of the ground before taking every step and then make lightning-speed adjustments such as lifting our foot to the right height, shifting our weight forward, changing our center of gravity and gait, and determining the force required to launch off of one foot and successfully land on the other. Every step requires a fast collection of data followed by fast processing, rapid problem-solving, and quick-fire action, followed by another round of data collection, processing, problem-solving, and adjustment, and so on. Incredibly, our brains do all of these calculations without ever once stopping to consciously think about it.

It's a wonder we don't require two full minutes between steps.

From an evolutionary standpoint, walking on uneven surfaces activates a closed loop system in the human brain that developed when we became bipedal, around five million years ago. With this evolutionary leap, our brains began evolving at an unprecedented speed. So it follows that over millions of years, we've developed the apparatus necessary to process the colossal amount of data required to make us highly skilled bipedal organisms. We have been perfecting this talent for a long time and at a very high price.

Even today, the benefits we receive from walking on an uneven terrain are astonishing: improvements in equilibrium, spatial orientation, memory, focus, reaction time, and overall cognitive fitness. Real time sensory input from our feet and eyes force the brain to make billions of calculations in milliseconds, and this turns out to be similar to exercising every area of the brain all at once.

So, one of the best workouts we can give our brains is walking rapidly on uneven surfaces. It's the equivalent of taking our brains to the gym to lift weights all day long.

In a controversial study, Dr. Arthur Kramer, a professor at the University of Illinois at Urbana, studied the effects walking had on the cognitive abilities of senior citizens. After six months of walking for short periods each day, Kramer measured significant improvements in both memory and attention. Though uneven surfaces are more ideal than even ones, there is now evidence that walking on *any* surface has cognitive benefits beyond just encouraging blood flow to the body and brain. According to Dr. Michael Merzenich, the relationship between movement and cognition cannot be separated because "movement is inextricably controlled on the basis of 'feedback' from our bodies and brains." This simply means that our brains turn into expeditious calculators as we quickly move over uneven surfaces.

The link between the locomotion of our bodies and how we perceive and process data is undeniable. Although this connection may have been forged millions of years ago when man stood upright, the cognitive benefits of walking are still as real today as they were for our earliest ancestors. Today, we know that walking not only leads to wellness, but it offers a wellspring of wisdom as well.

The Power of New

There is unanimous agreement among neuroscientists and psychologists that the human brain operates best when it is regularly subjected to new challenges. We have recently discovered that the brain benefits from a broad variety of problem-solving activities such as crossword puzzles and Sudoku. There also appear to be benefits when we mix these activities up: doing crosswords puzzles for a while and then switching over to Sudoku, and later, back again. The same goes for changing daily routines: trying a new route to work, a new sport, or a new hobby. Whenever we concentrate on learning something— anything—new, the brain activates neurotransmitters.

It doesn't matter what the new activity is—whether we are learning to bake a pie from scratch or tackling calculus—in each case the brain requires neurotransmitters to carry and store new information. This is because whenever we learn something new we are burning new biological "circuits" in the brain that challenge the old circuits we have relied on over and over and over again. Suddenly, we are creating more options and more pathways for the brain to select from.

Over time, searching for those new circuits becomes the new "normal"; the brain gets accustomed to creating new circuits for its own benefit. Successful learning has a certain momentum associated with it: The more the brain learns, the more it wants to learn. The desire to learn becomes habitual.

In this respect, brain fitness has a lot in common with physical fitness. The more we exercise, the more morphine-like chemicals are released in the brain. This positive feedback loop makes us want to exercise even more.

But when we exercise the same muscles over and over again, the muscles we are ignoring eventually atrophy from lack of use. That's the reason experienced physical trainers regularly change up workout routines. They want to give *all* the muscles in the body an opportunity to work so that our overall conditioning is better—not just a few areas.

The same goes for our brain. When we regularly mix up our workout by taking on new challenges, we benefit different areas of our brain. Like overall physical fitness, the more areas we workout, the better the result. This is one of the reasons that carefully designed brain fitness programs are so effective.

Need some easy ways to mix it up?

Mike Logan, an education counselor at Illinois State University, offers some easy suggestions:

> If you're right-handed, use your left hand for daily activities (or vice versa). Start with brushing your teeth left-handed, and practice until you have perfected it. Then try to build your way up to more complex tasks, such as eating. Changing simple activities drives our brain to make positive changes. Think of millions of neurons learning new tricks as you finally establish better control of that other hand!

In terms of businesses, social organizations, and governments, again variety paves the way for insightful thinking.

The best example of a government organization on the forefront of understanding the relationship between variety and cognition comes from the last place anyone would suspect: the U.S. Post Office.

The U.S. Post Office is a highly efficient organization that processes millions and millions of discrete items every day for mere pennies. Each job is highly regimented and routine. Yet, surprisingly, turnover is low and error rates are virtually nonexistent on a per transaction basis.

The post office may not be a bastion for innovative thinking, but they are remarkably sensitive to the effects that routine has on the human brain. Research shows that highly repetitive jobs such as sorting and delivering mail can quickly result in boredom, errors, poor morale, and high turnover among employees.

But over a decade ago the U.S. Post Office developed an innovative way to introduce variety in order to cut down on the risks associated with routine. Because postal tasks don't significantly vary from one location to the next, this presented a unique opportunity: Employees could exchange jobs with postal workers living in other parts of the country for short periods of time with virtually no disruption in productivity.

In addition to avoiding all the pitfalls associated with boredom, the program provided a second benefit: a marketing tool for recruitment. Job exchange quickly became one of the perks of working for the postal service: the prospect of travel and variety without sacrificing job security. What's more, according to the employees who took advantage of the program, changing locations was a great deal of fun. Although the tasks they performed remained exactly the same, they were introduced to new experiences in every area outside of their jobs: new co-workers, new neighborhood, new grocery store, and so on. The exchange program was the perfect solution to a tiresome, highly structured, and repetitious job, one that otherwise had all the makings of massive errors and unstoppable turnover.

A few large corporations offer similar programs, but for the most part, as jobs become increasingly specialized, many companies stubbornly cling to the notion that repetition leads to greater productivity. Given the evidence, it's unclear why this meme persists. If insightful thinking depends on making new connections in the brain, then varying routine and minimizing uniformity not only keep employees engaged but also coax insightful thinking to the forefront.

As Mike Logan points out, the opportunity to integrate more variety and learning into our everyday life exists no matter what kind of job we have. It may be as simple as using our left hand instead of our right one, using one new vocabulary word throughout the day, working while standing at our desk, or doing a Sudoku puzzle at breakfast. Variety stimulates new circuits, and new circuitry is what insights are all about.

Training, Braining, and Gaining

Brain fitness may be a newly emerging field, but there is no longer debate about whether products designed by neuroscientists such as Dr. Michael Merzenich provide a cognitive advantage. Today, these products are being adopted by a new breed of brain fitness gyms around the world as well as a growing number of public and private schools.

In fact, the Japanese government is so convinced brain fitness is an essential weapon against growing complexity, they are the first country to invest $350 million into new cognitive tools designed to arm the next generation. Japanese leaders not only believe brain fitness will give their citizens a cognitive edge in the world economy, they also see it as an essential safeguard against Alzheimer's, dementia, and other cognitive diseases associated with aging.

The Japanese government gets it.

Preparing the human mind for complex problem-solving offers the single greatest socioeconomic advantage possible across all fields and all industries. In this respect, the Japanese government views improvements in cognition as a long-term strategic advantage and health mandate. With so many studies showing a direct relationship between twenty to sixty minutes of brain fitness a day and improvements in health, problem-solving, spatial orientation, memory, reaction time, and sharper focus, giving the brain a daily workout is—in the words of a colleague—"a no-brainer."

Bring Back the Break

A great deal of research has been conducted on the benefits of periodically stepping away from a problem in order to look at it from a fresh perspective. One of the interesting aspects of insightful thinking is that people who have experienced sudden breakthroughs all report feeling "stuck" just prior to their "aha!" moment. They bump up against some cognitive barrier that prevents them from making the new connections needed to solve the problem. But once they step away—relax and temporarily think about something else—bam! The answer strikes from out of nowhere.

Today, many studies show the important relationship between unstructured "play" and cognition among schoolchildren. In a 1977 study, Saltz, Dixon, and Johnson demonstrated that playtime had a dramatic effect on intelligence tests because it encouraged creativity, vocabulary use, and spontaneous problem-solving. More recent research conducted by Dr. Anthony Pellegrini, psychology professor at the University of Minnesota, and Robyn Holmes of Monmouth University reveals the positive effect recess has on learning: "Children are less attentive to classroom tasks during longer, compared to shorter, seatwork periods."

Pellegrini and Holmes's research reveals that children learn and perform better in school immediately following recess. He also indicates that short periods of concentrated study work better than long periods without breaks. In similar studies, Japan and Taiwan report that instruction is more successful when the learning periods are "relatively short" and "intense" and "there are frequent breaks between these work periods." In fact, in China it has become standard practice to provide unstructured play in the school system every fifty minutes to facilitate concentration and learning in the classroom.

Moreover, the benefits of recess don't just apply to children who are loading content and learning to solve problems. Many studies confirm that the same benefits occur when frequent breaks are scheduled in the workplace, especially when an employee's work requires long periods of focused attention.

So it comes as no surprise that there appears to be a relationship between taking breaks, relaxation, and insightful thinking.

It turns out there's a reason we have some of our best ideas when we wake up in the morning or when we are taking a shower, driving, or just sitting quietly. Often, when we return from a short vacation, problems that once seemed intractable suddenly become solvable. How many times have we felt stuck, then stood up, stretched our legs, and walked away only to come sit back down and discover what we were doing wrong?

Stories such as the spilled can of paint that led to Jackson Pollock's landmark innovation in American abstract art or Nobel physicist Richard Feynman's habit of madly scribbling notes on cocktail napkins while sitting in topless bars all have one thing in common. In each instance, the insightful breakthrough occurred while the person was relaxing and allowing his mind to wander, providing it freedom to make new connections and uncover hidden solutions.

We have discovered that the human brain is much more likely to have a breakthrough when it is producing alpha waves—the kind of waves that occur when we are meditating. "The relaxation phase is crucial," says cognitive neuroscientist from Northwestern University Dr. Mark Jung-Beeman, noting that some of our best thinking is done while we're half asleep.

According to Dr. Joydeep Bhattacharya, psychologist at Goldsmiths University of London, one of the predictors of insightful thinking is the appearance of alpha waves in the right hemisphere of the brain. These alpha waves allow the human brain to respond to new ideas and information by encouraging the mind to "wander." Thanks to sophisticated measurement devices, Dr. Bhattacharya is now able to predict when insightful thinking will take place, having found that participants solved insight puzzles "several (up to 8) seconds before the behavorial response."

Insight requires the mind to be in a relaxed state and in a good mood. Dr. Karuna Subramaniam of the University of California at San Francisco, along with Kounios, Parrish, and Jung-Beeman, discovered the following: "Participants higher in positive mood solved more problems, and specifically more with insight, compared with participants lower in positive mood. . . . Positive mood alters preparatory activity in the ACC [anterior cingulate cortex], biasing participants to engage in processing conducive to insight solving."

What are of these scientists trying to tell us? We often become our own worst enemy when it comes to deploying insightful thinking.

It's a lot like trying to fall asleep when we know we have a big presentation the next morning. As more time passes and we have less and less time for sleep, the more anxious we become as we try to *force* ourselves to relax and sleep. This same sort of conundrum applies to insight. Trying to force ourselves to have an insight doesn't work. In fact, the more we try, the further from a relaxed state we get and the less likely insight is to appear.

Pressure, stress, judgment, negative attitudes, and bad moods all inhibit insightful problem-solving. Because these orientations are within our control, we can increase our capacity to manage complexity by simply relaxing the mind and creating a positive, upbeat mood so insights can be coaxed out of hiding. This may mean taking on complex issues just after resting, yoga, meditation, or a short walk outside. It may mean listening to music, taking a warm bath, or just sitting quietly. Anything we can do to "unwind" and allow the mind to relax is good for insight.

Damaging Distractions

Research has demonstrated that the brain "prepares" itself milliseconds before it uses insight to solve a problem. This preparation includes a deliberate effort by the brain to shut out both internal and external distractions so that it can wander in a highly focused way and thereby make *meaningful* new associations. As opposed to creative thought that may or may not be productive, insight is an accurate method of solving complex problems. As a result, during insight the brain wants to do two seemingly contradictory things: freely search for new connections while also ruthlessly eliminating options that don't work.

In addition to relaxing, insight is facilitated when extraneous thoughts about prior and future events are eliminated, creating what Dr. John Kounios, professor at Drexel University, calls a "clean slate."

In his 2006 paper, "The Prepared Mind," Kounios wrote the "ACC [anterior cingulate cortex] may be involved in suppressing irrelevant thoughts, such as daydreams or thoughts related to the preceding trial, thus allowing solvers to attack the next problem with a 'clean slate.' This explanation assumes that insight processing is more susceptible to internal interference than is noninsight processing, thereby necessitating greater suppression of extraneous thoughts."

To this end, meditation and other activities that help to bring the mind into the "present" increase opportunities for insightful problem-solving.

On the one hand, a person might argue that Wag Dodge and John Saar were both contending with a huge distraction when they had their insights. In both cases a blazing inferno was quickly approaching. On the other hand, the impending disaster may have forced both men to focus on their problem to the exclusion of all other thoughts. It's safe to say that when our life is in peril, other concerns fall by the wayside, allowing for extreme concentration.

The role danger plays in compelling insight is unknown, but based on what neuroscientists have been able to observe in the laboratory, insight follows a period of extreme focus and concentration. So whether or not we face danger or carve out quiet, undistracted time to think, clearing the deck appears to be a prerequisite to insight.

Collaborating with Complexity

One of the surefire ways to slow down complexity is to stop buying and doing more. When we willingly choose to keep adding new things to our lives—more goals, activities, needs, wishes, products, and so on—we become unconscious accomplices of complexity.

For example, when we purchase a second vacation home, it rarely occurs to us that we are also doubling the number of toilets that can leak, yards that have to be watered, bills that have to be paid. When we buy a second car, it's the same thing: twice the oil changes, tire replacements, insurance payments, and so on. How about that ministorage where we are saving all those things we might need someday? Or those overstuffed cabinets in our garage, our children's closets, and even our desk drawers? Now consider the number of afterschool activities our children attend and the amount of homework they are required to do every day. Add in two jobs to keep up with the bills while also trying to set money aside for college and retirement and a family vacation. Our lives have become so busy, so congested, so complicated, that it's no wonder we have no time to think.

Time to slow the train down.

Just because we have an ever-growing selection of products and technologies, opportunities, and offers available to us doesn't mean we have to rush into them. In fact, an argument can be made that *too much* contributes just as quickly to gridlock as *not enough*.

In an environment where there are too many choices, too many decisions, too much information, and too many demands on our cognition, it pays to be judicious about the complexity we voluntarily sign up for. When we make the decision to streamline our lives, we also create time and room to think with focus and intent. In a complex world, time to stand back and look at the big picture, time to consider our options more carefully, time to make more deliberate decisions, and time to breathe are necessities for survival.

Eating, Sleeping, and Exercising

Over time we've learned that diet, rest, and exercise play a crucial role in almost all areas of life, so it's not surprising that we have discovered these habits have an impact on improving insight.

Let's begin with diet.

Are some foods better for the brain than others?

Recent research seems to be headed in that direction. Our diet acts just like the quality of gas we put into our car. Cheap fuel can gunk up the works when it comes to blood flow, slowing down the quantity of blood to the brain. What's more, the performance of the neurotransmitters in the brain depends on what we eat because neurotransmitters are made from amino acids, and amino acids come from the protein in our diet (meat, fish, and cheese).

So, the first priority for getting the highest brain performance is to consume sufficient amounts of protein each day. Insights are cognitively taxing to the human brain, so it makes sense that fueling our neurotransmitters with high-octane fuel—protein—is essential for high-powered thinking.

Next come antioxidants from foods like blueberries, Matcha green tea, and walnuts, which stave off cognitive cell damage. We need healthy cells in order to burn new circuitry, and as we established earlier, the more circuits we burn, the more fodder we have for insightful problem-solving.

In addition, high-quality dark chocolate is known to activate the production of dopamine, which has a direct impact on both memory and learning, and so this may also give the brain the boost it needs to encourage insight.

Finally, Omega-3 fatty acids from fish (coldwater fish like wild salmon) have been proven to suppress brain cell inflammation. The human brain is more than 60 percent fat, so an insufficiency of Omega-3 fats leads to depression, poor memory, and learning disabilities—all known obstacles to insight.

Many new books now suggest that certain foods are better for brain health than others. And every year scientists and nutritionists learn more about the ideal mix of fuels required for thinking. But so far the fundamentals appear remarkably consistent no matter which expert you believe: Keep our protein, antioxidants, and Omega-3 acids high; treat ourselves to high-quality chocolate every once in a while; remember to take a multivitamin (loaded with B-complex and potassium) every day; and drink sufficient amounts of water.

How about exercise?

Again, the impact of physical health plays a large role in the brain's readiness to take on complex problems. Children who spend all day sitting in front of a computer, television, or video game aren't getting anywhere near the cognitive workout that walking on uneven surfaces, playing outdoors, and interacting with other children provide.

Foods Associated with Higher Brain Functioning

Blueberries, Cranberries, Raspberries, Strawberries, and Blackberries

Avocados, Carrots, Tomatoes, and Eggplant

Spinach, Kale, Romaine Lettuce, Brussels Sprout, and Broccoli

Wild Salmon, Sardines, and Herring

Sunflower Seeds, Sesame Seeds, Flax Seeds, and Pumpkin Seeds

Walnuts, Hazelnuts, Brazil Nuts, Filberts, Almonds, Cashews, and Peanuts

Oatmeal

Brown Rice

Whole Grain Breads

Popcorn

Barley

Black Beans, Lima Beans, Pinto Beans, Kidney Beans, Garbanzo Beans, Great Northern Beans, Navy Beans, Peas, and Lentils

Cantaloupe, Mangos, Apricots, Papaya, Oranges, Prunes, Red Grapes, and Cherries

Freshly Brewed Green Tea

Dark Chocolate

Water (eight glasses per day)

Sage

Eggs

Curry Spice

Research shows that being sedentary and repeating a narrow range of tasks over and over again mean depending on the same circuits in the brain instead of creating new ones. So in addition to these activities lowering the blood supply to the brain, children become less familiar with burning new circuits and thinking in innovative ways.

Dr. Michael Merzenich describes the harm that the absence of variety causes the human brain this way: "The computer-game freak or couch potato is in a comparative motor learning and complex thinking rut when it comes to the elaboration of response to information the brain is seeing."

The first and obvious reason exercise has an impact on insight is that it increases blood flow to the brain. Research indicates regular exercise facilitates generating new brain cells and has positive influences on areas responsible for learning and memory (hippocampus).

In short, when we improve blood flow through exercise, we are also improving idea flow. Blood flow is so crucial to the functioning of the human brain that scientists are on the cusp of discovering that increasing blood flow may reverse the effects of aging in both the body and the mind. According to Dr. Michael Merzenich, in a preliminary study at the University of California at San Francisco, there is promising evidence that older rats that have lost their physical agility to climb ropes and are suffering the early onset of various cognitive diseases have reversed both conditions when blood flow increased to their brains. Suddenly, they were able to move like much younger rats, and problems associated with early dementia, in some cases, reversed themselves.

But *cognitive fitness requires* new *learning to take place, so physical exercise that incorporates new sensory experiences is the best way to give both our bodies and brain a workout.* That means running around the neighborhood or mountain biking down a bumpy trail offers more cognitive benefits than simply walking on a treadmill day after day. It also means that the corporate gym doesn't cut it anymore. Going outside and walking or playing a quick round of badminton or basketball is cognitively more challenging and bound to get new neurotransmitters activated in a way that sets the stage for insightful thinking.

Finally, we come to a topic everyone is talking about these days: sleep.

In a 2002 study it was reported that 75 percent of Americans have trouble sleeping several nights a week, and 37 percent admit this interferes in their daily lives. This may account for the number of sleep disturbance clinics that have exploded throughout the United States in just a single decade.

Sleep deprivation has a profound effect on the frontal lobe, the area responsible for complex problem-solving and insight. Without rest, the brain has no time to consolidate information, so learning becomes extremely difficult. This is just another way of saying what we already know: When we don't get enough sleep it becomes impossible to concentrate. Our problem-solving abilities greatly diminish.

In summary, adequate rest (six to eight hours each night), plenty of brain food, and physical exercise have all been shown to improve complex cognitive processing. So, if we want to improve our odds for insightful thinking, we need to give our brains all the fuel, rest, and exercise they require.

Swimming in a Clear Pond

The impact of altered brain chemistry on insight has not been adequately studied in order to form any reasonable conclusions. So, here I can only err on the side of caution.

It is widely known that antidepressants are now the most prescribed medication in the United States. According to the Centers for Disease Control and Prevention, approximately 118 million prescriptions for antidepressants were issued in the United States in 2005. Between 1995 and 2002, it is estimated that the use of antidepressants skyrocketed by 48 percent.

Regrettably, this trend is similar among most industrialized nations. Mood-altering drugs have become a quietly sanctioned international epidemic.

But pharmaceuticals are just one form of medication, and they aren't the only culprits responsible for altering the chemistry in the brain. We medicate with junk food, alcohol, exercise, marijuana, television, and, sometimes, even work. All of these forms of medication have an impact on brain chemistry. Eat a doughnut, and the brain gets high on sugar and fat. Have a beer, and it responds to the alcohol.

Stand next to someone smoking, and the brain gets a hit off of second-hand nicotine.

It's easy to prove what alcohol or tobacco does to brain chemistry, but it's much harder to prove how long periods of television-viewing or too many hours at work affect our problem-solving abilities. The effect is more nebulous. But regardless of how we choose to medicate, one thing is certain: *We are relying on an artificial means to alter our mood and orientation to our environment.*

To this end, it's important to better understand how antidepressants and other popular forms of medication are affecting the evolution of the human brain. For example, in the last decade depression has been linked to a chemical imbalance in the brain. When we feel depressed or anxious, chemicals called dopamine, seratonin, and norepinephrine, which act as neurotransmitters in the brain, are no longer in balance. Antidepressants are designed to restore the chemical balance so that our moods improve.

Because there is growing evidence that being in a positive mood *increases* the probability of insightful thinking, we might expect a new generation of happy pills to produce more breakthrough "aha!" moments. But oddly enough, the increase in chemical neurotransmitters caused by antidepressants appears to have almost no effect on increasing frontal lobe effectiveness. This means that chemicals designed to make us feel happier and more relaxed do not appear to be doing anything to *improve* our ability to solve complex problems or stimulate insights.

So if antidepressants aren't facilitating insights, this leaves only two possibilities: Changing our brain chemistry inhibits the evolution of insight, or it has no effect at all.

Until we have clinical evidence either way, my suggestion is to try, whenever possible, to resist the temptation to artificially modify the chemistry of the brain—no matter whether it's alcohol, marijuana, antidepressants, or even aspirin. Until the impact of medications on the development of complex thinking is fully understood, why not take a "better safe than sorry" approach?

That said, what are the natural alternatives to medicating with cocktails, cigarettes, work, and prescription drugs?

When it comes to depression, there is growing evidence that many mild, seasonal, or episodic cases can be successfully managed by taking a few simple steps. For example, exercise is known to produce endorphins in the brain. Clinical studies indicate that a large percentage of people diagnosed as depressed feel happier and more motivated once they begin to exercise regularly. The same kinds of results have been reported when a person's social interactions increase and also after a person begins talk therapy.

Although there is no way to know how many people diagnosed as clinically depressed can be effectively cured without taking prescriptive drugs, it's difficult to believe that one quarter of the population in the United States requires a prescription to alter the chemistry of their brains. For many people, Mother Nature offers an equally powerful, lasting remedy. When we respect the brain's ability to make chemical adjustments of its own, we are allowing it to continue to evolve without interference—good, bad, or indifferent.

Imagine

Just for a moment, imagine a world where government leaders pared their cabinets down to a small group of four to nine people; where farmers were subsidized to grow "brain foods" so that they cost nothing to eat and junk foods became really expensive; where corporations outlawed "overtime" and paid employees to sleep seven hours every night. Imagine insurance companies paying clients to exercise, take an art class, or go back to school instead of paying for medications designed to manage blood pressure, depression, anxiety, diabetes. Imagine a world where every company had an old-fashioned "recess bell" signaling its employees to stop work, step away from their desks, and relax; where every day wasn't jam-packed with activities and every drawer and closet we owned wasn't stuffed to the brim; where walking on the dirt was preferable to driving a car and playing a brain fitness game once a day was all we needed to do to fend off dementia and a battery of other cognitive illnesses.

Today's neuroscientists are learning that such a world makes much more sense than the one we are living in. They have known for some time now that the key to the survival of the species is locked deep inside the human brain: What is good for cognition is not only good for the perpetuation of the species but also for other life-forms and the planet. The more we do to conquer the cognitive threshold, the more likely we are to successfully meet the complex challenges that lie ahead.

Cognition, Species, Planet. They are inextricably linked.

In ancient civilizations, the terms of their existence were dictated by the slow, deliberate hand of evolution. No matter how quickly they progressed, the evolution of their brains eventually fell behind. It was just a matter of time before the discrepancy caught up with them. However, today modern man has the ability to look inside the brain and finally understand how it works—we stand at the edge of making the most important discovery since humans became mammals, forged social groups, and discovered two-legged locomotion: *insight-on-demand.*

As we learn more about the power of this rare and gifted cognitive ability, we are learning that the gap between the uneven rate at which the human brain can evolve and the rapid pace at which we generate and discover complexity can be bridged. By coaxing insight out of the shadows to claim its rightful place in human history, modern man can surmount a cognitive limitation that has trapped humans in a repetitive pattern of ascension and collapse for millions of years. We can take the next step.

— 13 —

On the Threshold

So, if I understand you right, you're saying that if we parachuted a Neanderthal into Times Square today, he wouldn't do well."

"That's right."

"Likewise, if you put us in a time machine and sent us five million years into the future, our brains wouldn't stand a chance either."

"Probably not."

"All you're saying is that evolution is really, really slow, so eventually our brains fall behind."

"Yes."

"Kind of like Lucy trying to take the chocolates off the conveyor belt and get them into a box. The conveyor belt starts speeding up and suddenly she can't grab the chocolates fast enough."

"Well, sort of . . ."

"So things start unraveling quickly unless we figure out a way to slow down the conveyor belt, or grab the candy faster."

"Ideally we'd do both. Some of the chocolates are starting to fall on the floor."

ON THE MORNING OF August 29, 2004, I had an important insight.

I was on my way to the birth of my nephew, Ben, when I began thinking about what we really understand about life—not what we *believe*, what we have proven.

The first thing that came to mind was evolution.

For over a century now, scientists around the world have been unearthing physical evidence buried far beneath the planet's crust and deep in the molecular biology of DNA that proves that the principles of evolution govern all life, both large and small. *We now know with absolute certainty that every organism, from ants to elephants, marches to the laws of Natural Selection.*

Given this, it seems curious that whenever we speak about evolution, we speak about it in the past tense. We treat evolution as if it were something that happened to our prehistoric ancestors. Yet, the fact is that we are all evolving right now—right this minute—right as I write these words.

Why do we have such a difficult time integrating evolution into the present? When did talking about where we stand in the continuum of biological change become a controversial topic? *How did a fundamental tenet in science become synonymous with godlessness? Determinism? Liberalism?*

Ever since the publication of Charles Darwin's *On the Origin of Species*, we have had an ambivalent relationship with evolution. The vast majority of us are willing to acknowledge its importance in the past, but we don't seem to think it is relevant to our problems today. It seems strange that we don't see any connection between the single most important principle governing all life on earth and the massively complex problems that have stalked us for multiple generations such as global warming, terrorism, poverty, pandemic viruses, nuclear threat, a global financial crisis, and declining public education.

Why is that?

Is it possible we have marginalized evolution because once we acknowledge the fact that we are still evolving, then we also have to accept the fact that there's *no way for humans to progress faster than evolution will permit?* That would mean that human beings have biological constraints that limit the kinds of problems we can solve.

Biological constraints?

Nobody wants to hear that.

Even so, that doesn't change the facts. At some point, the complexity and magnitude of the problems a civilization must solve simply exceed our biological capabilities. The point at which complexity and evolution collide is the *cognitive threshold,* and it has been the trigger for the demise of every advanced civilization since the beginning of humankind.

This is a daunting realization.

But I admit that while coming to the realization, I never once thought of Lucy on the assembly line.

Yet, it's the perfect metaphor.

When we can't develop new cognitive capabilities fast enough, any number of irrational behaviors are bound to materialize.

First, we start stuffing chocolates in our mouth and shirt, trying to hide the problem. But the conveyor belt keeps speeding up, and the candy just keeps coming.

Then, when we can't eat or hide the chocolates anymore, we freeze. We become gridlocked, as chocolates start falling faster and faster to the ground.

Eventually, we are forced to stop the conveyor belt and shut the factory down. With any luck, a little later we reorganize, open under new management, and start over again. Then things go fine again— for a while. But the conveyor belt starts picking up speed again and . . .

Nature has provided an elegant solution to our conundrum: an astonishing problem-solving capability buried deep in the human brain called *insight.* The recent discovery of insight is no less remarkable than suddenly realizing that all we have to do is hold the box at the end of the conveyor belt and let the chocolates drop in, one by one.

So no matter how fast the belt gets going, keeping up is easy.

Today, for the first time in history neuroscientists have the ability to look inside the human brain and witness insight's true talents. They acknowledge that insight is spontaneous and rare, but it is also the only known cognitive antidote to complexity.

But in addition to discovering insight, we are also the first civilization to have the knowledge, technology, and resources to stop the pattern. No other civilization before us had such command over its environment and so many tools at its disposal to alter the course of human ascension and collapse. And, therefore, no other civilization has had so much cause for optimism.

Here, visionary Paul Hawken gets the final word:

"The poet Adrienne Rich wrote, 'So much has been destroyed I have cast my lot with those who, age after age, perversely, with no extraordinary power, reconstitute the world.' There could be no better description. Humanity is coalescing. It is reconstituting the world, and the action is taking place in schoolrooms, farms, jungles, villages, campuses, companies, refugee camps, deserts, fisheries, and slums."

Armed with our new understanding of collapse, will modern civilization allow evolution to take its rightful place in the twenty-first century? Will we restore the balance of knowledge and beliefs and allow uncommon insights to cure what ails humanity? Will we answer the clarion call of the Watchman's Rattle in the dead of night?

Acknowledgments

In 1676, Isaac Newton wrote in a letter to Robert Hooke, "If I have seen a little further it is by standing on the shoulders of Giants." I, too, have been generously lifted onto the shoulders of minds far greater than my own.

I begin by giving my deepest gratitude for the work of Charles Darwin, and Drs. E. O. Wilson, James Watson, and Richard Dawkins, without which *The Watchman's Rattle* would not have been possible. Their insights in evolution, sociobiology, consilience, and meme theory were the dots—my only contribution was to connect them.

I want to thank Dr. Michael Merzenich, Dr. Steven Chu, Dr. Charles Townes, Dr. Philip Brownell, Dr. John Ratey, and Dr. John Sumser for their counsel; Leon and Sylvia Panetta at the Panetta Institute and Neil Patterson at the E. O. Wilson Biodiversity Foundation; and Drs. John Kounios and Mark Jung-Beeman, whose groundbreaking research in the field of insight provides great optimism for our future.

I will never be able to repay my debt to Arthur Klebanoff, the president of the Scott Meredith Agency, whose vision, business acumen, and leadership are largely responsible for the book's success. I owe a similar debt to the team at Vanguard Press and Perseus Books Group, most notably Roger Cooper and Rick Joyce, for courageously embracing the ideas in the book and investing the resources needed to bring them to life.

A special thanks to two of the greatest editors a writer could hope to have in her corner, Dana Benningfield, and to Christine Marra, whose editorial experience was indispensable. I am also deeply grateful to Sandi Mendelson and David Nelson for crafting the message and tirelessly shaking the rattle so that the world would take notice.

I want to thank my family, Sam, Mat, Michael, and Shelly, for their unwavering love. And my deep gratitude goes out to my friends who never had a moment's doubt: Brigitte, Jane, Rosy, Kitty, Joy, Francine, Cheryl, Mary, and Sharon. A special thanks to Michael Geary for helping in Afghanistan when I couldn't go.

And last but not least, I thank my trusted companion, Tonka. The slow sound of your breath as you slept under my desk always brought me back down to earth.

Further Reading

The Watchman's Rattle is a convergence of history, biology, neuroscience, meme theory, and the study of complexity. Therefore, more than specific books, I feel that a breadth of reading is important. The first place to begin is to read Charles Darwins's *original* works so there is a good foundation on which to build.

Bar-Yam, Yaneer. *Making Things Work.* Cambridge, MA: Knowledge Press, 2004.

Blackmore, Susan. *The Meme Machine.* New York: Oxford University Press, 1999.

Brodie, Richard. *The Virus of the Mind.* Seattle, WA: Integral Press, 2009.

Carter, Rita. *The Human Brain Book.* New York: DK Publishing, 2009.

Darwin, Charles. *The Descent of Man.* New York: The Penguin Group, 2007.

———. *The Expression of the Emotions in Man and Animals.* New York: Oxford University Press, 1998.

———. *On the Origin of Species.* New York: Sterling Publishing, 2008.

———. *The Voyage of the Beagle.* Washington, D.C.: National Geographic Society, 2009.

Dawkins, Richard. *The Extended Phenotype.* New York: Oxford University Press, 1999.

———. *The Selfish Gene.* New York: Oxford University Press, 2006.

Diamond, Jared. *Collapse.* New York: Viking Penguin, The Penguin Group, 2005.

Diamond, Jared. *Guns, Germs, and Steel.* New York: W. W. Norton & Company, 1999.

Hawkins, Jeff. *On Intelligence.* New York: Times Books, 2004.

Lilla, Mark. *The Stillborn God: Religion, Politics, and the Modern West.* New York: Alfred A. Knopf, Random House, 2007.

Mann, Charles C. *1491.* New York: Vintage Books, Random House, 2006.

Pinker, Steven. *The Stuff of Thought.* New York: The Penguin Group, 2007.

Rousseau, Jean-Jacques. *The Social Contract.* Translated by Maurice Cranston. New York: Penguin Group, Penguin Classics Various Editions, 1968–2007.

Schwartz, Glenn M., and John J. Nichols. *After Collapse.* Tucson: The University of Arizona Press, 2006.

Sweeney, Michael S. *Brain: The Complete Mind.* Washington, D.C.: National Geographic Society, 2009.

Toffler, Alvin. *Future Shock.* New York: Bantam Books, Random House, 1984.

Wright, Robert. *The Moral Animal.* New York: Vintage Books, Random House, 1994.

Wilson, Edward O. *Consilience.* New York: Alfred A. Knopf, 1998.

———. *The Future of Life.* New York: Vintage Books, Random House, 2002.

———. *On Human Nature.* Cambridge, MA: Harvard University Press, 2004.

———. *Sociobiology.* Cambridge, MA: Belknap Press of Harvard University Press, 2000.

Notes

Introduction

p. xiii "It's dangerous to state the obvious"
Interview with Dr. E. O. Wilson, Harvard University, Cambridge, MA, July 1, 2009.

p. xiii The first piece of the puzzle arrived in 1859
Charles Darwin, *On the Origin of Species* (New York: Sterling, 2008).

p. xiv Then in 1953 came the discovery by James Watson and Francis Crick
James D. Watson, *The Double Helix* (New York: Touchstone, 2001).
Robert C. Olby, *The Path to the Double Helix: The Discovery of DNA* (Mineola, NY: Dover, 1994).

p. xiv By the time E. O. Wilson's controversial book *Sociobiology*
Edward O. Wilson, *Sociobiology: The New Synthesis* (Cambridge, MA: Belknap Press, 2000).

p. xiv Then one year later, Richard Dawkins released his book *The Selfish Gene*
Richard Dawkins, *The Selfish Gene* (New York: Oxford University Press, 2006).

p. xv The first company to offer me a job
David E. Weisberg, *The Engineering Design Revolution* (Englewood, CO: www.cadhistory. net, 2008).
Peter Petre, "How GE Bobbled the Factory of the Future," *Fortune*, November 11, 1985.

Chapter 1.
A Pattern of Complexity and Collapse: Why Civilizations Spiral

p. 2 In his book *Making Things Work*
Yaneer Bar-Yam, *Making Things Work* (Cambridge, MA: Knowledge Press, 2004).
U.S. Department of State Web site: www.state.gov/s/c+/other/des/1230085.htm.

p. 3 Between 2600 BC and 900 AD, the highly advanced Mayan civilization
Matthew Markowitz, "The Mayans, Climate Change, and Conflict," ICE Case Studies, No. 112 (2003), http://www1.american.edu/ted/ice/maya.htm.
Charles C. Mann, *1491* (New York: Vintage, 2006).
Glenn Welker, "Mayan Civilization," http://www.indians.org/welker/maya.htm.
Patrick L. Barry, "The Rise and Fall of the Mayan Empire," January 6, 2001, http://www.firstscience.com/home/articles/origins/the-rise-and-fall-of-the-mayan-empire_1387.html.
David L. Webster, *The Fall of the Ancient Maya* (New York: Thames and Hudson, 2002).

Michael Lemonick, "Mysteries of the Mayans," http://www.indians.org/welker/maya.htm.

"Collapse: Why Do Civilizations Fail?" http://www.learner.org/interactives/collapse/mayans.html.

Stefan Lovegren, "Climate Change Killed Off Mayan Civilization, Study Says," *National Geographic News*, March 13, 2003.

John Ness, "Fall of the Mayan," *Newsweek*, March 24, 2003.

Elin C. Danien and Robert J. Sharer, *New Theories on the Ancient Maya* (Philadelphia: University Museum, University of Pennsylvania, 1992).

Joseph Tainter, *The Collapse of Complex Societies* (Cambridge: Cambridge University Press, 1990).

David Freidel and Linda Schele, *A Forest of Kings: The Untold Story of the Ancient Maya* (New York: HarperPerennial, 1992).

Heather Irene McKillop, *The Ancient Maya: New Perspectives* (New York: Norton, 2006).

"Mayan History," http://www.crystalinks.com/mayanhistory.html.

p. 4 Professor Gerald Haug is the best-known proponent

Steve Connor, "How Drought Helped Drive a Long-Lost Civilization to Extinction," *The Independent*, March 14, 2003.

p. 4 According to Michael D. Lemonick

Michael Lemonick, "Mysteries of the Mayans," http://www.indians.org/welker/maya.htm.

p. 4 "The Maya were obsessed with war"

Michael D. Coe, *Breaking the Mayan Code* (London: Thames and Hudson, 1992).

p. 4 But according to Jared Diamond

Jared Diamond, *Collapse* (New York: Viking Penguin, 2005).

p. 8 John Stanton, CBS, ABC and CNN commentator

John Stanton, "Evolutionary Cognitive Neuroscience," *Dissident Voice*, June 30, 2007.

p. 8 What's more, most of us agree that the brain will continue to evolve

Nicholas Wade, "Researchers Say Human Brain Is Still Evolving," *New York Times*, September 8, 2005.

William H. Calvin, Terrence Deacon, Ralph L. Holloway, Richard G. Klein, Steven Pinker, John Tooby, Endel Tulving, and Ajit Varki, "The Evolution of the Human Brain" (Center for Human Evolution, proceedings of Workshop 5, Bellevue, Washington, March 19–20, 2005), http://www.futurefoundation.org/programs/che_wrk5.htm.

"Human Brain Evolution Was a 'Special Event,'" *HHMI Research News*, December 29, 2004, http://www.hhmi.org/news/pdf/lahn3.pdf.

Jane Bradbury, "Molecular Insights into Human Brain Evolution," *PLoS Biology* 3(3) (2005).

Kate Melville, "Evolution of the Human Brain Unique," Scienceagogo, December 29, 2004, http://www.scienceagogo.com/news/20041129182724data_trunc_sys.shtml.

George F. Striedter, *Principles of Brain Evolution* (Sunderland, MA: Sinauer Associates, 2004).

Christopher Willis, *The Runaway Brain: The Evolution of Human Uniqueness* (New York: Basic Books, 1994).

Mihail C. Roco and Carlo D. Montemagno, *The Coevolution of Human Potential and Converging Technologies* (New York: New York Academy of Science, 2004).

Robert J. Sternberg and Janet E. Davidson, *The Nature of Insight* (Cambridge, MA: MIT Press, 1995).

p. 11 Once again, Dr. Yaneer Bar-Yam sheds light
Yaneer Bar-Yam, *Making Things Work* (Cambridge, MA: Knowledge Press, 2004).
p. 11 Complex problems that cannot be solved eventually manifest themselves
A. W. Kruglanski and D. M. Webster, "Motivated Closing of the Mind: 'Seizing' and 'Freezing,'" *Psychological Review* (1996).
W. Kruglanski, D. M. Webster, and A. Klem, "Motivated Resistance and Openness to Persuasion in the Presence or Absence of Prior Information," *Journal of Personality and Social Psychology* 65(5) (November 1993).
M. Mitchell Waldrop, *Complexity: The Emerging Science at the Edge of Order and Chaos* (New York: Simon and Schuster, 1992).
Melanie Mitchell, *Complexity: A Guided Tour* (New York: Oxford University Press, 2009).
John J. Miller and Scott E. Page, *Complex Adaptive Systems: An Introduction to Computational Models of Social Life* (Princeton, NJ: Princeton University Press, 2007).
Robert Geyer and Samir Rihani, *Complexity and Public Policy: A New Approach to 21st Century Politics, Policy, and Society* (New York: Routledge, 2010).
Len Fisher, *The Perfect Swarm: The Science of Complexity in Everyday Life* (New York: Basic Books, 2009).
Linda Elder and Richard Paul, "Critical Thinking in a World of Accelerating Change and Complexity," *Social Education*, December 1, 2008.
Arthur Whimbey and Jack Lochhead, *Problem Solving and Comprehension* (Mahwah, NJ: Lawrence Erlbaum, 1999).
John H. Holland, *Hidden Order: How Adaptation Builds Complexity* (New York: Basic Books, 1996).
Bryan K. Hanks and Katheryn M. Linduff, *Social Complexity in Prehistoric Eurasia: Monuments, Metals, and Mobility* (New York: Cambridge University Press, 2009).
N. Jausovec and K. Bakracevic, *Creativity Research Journal* 8(1) (January 1995).
G. A. Miller, "The Magical Number Seven, Plus or Minus Two: Some Limits on Our Capacity for Processing Information," *Psychological Review* 63(2) (1956).
p. 11 Then, as conditions grow more desperate
R. E. Petty and J. A. Krosnick, *Attitude Strength: Antecedents and Consequences* (Mahwah, NJ: Lawrence Erlbaum, 1995).
A. R. Pratkanis, S. J. Breckler, and A. J. Greenwald, *Attitude Structure and Function* (Hillsdale, NJ: Lawrence Erlbaum, 1989).
S. Budner, "Intolerance of Ambiguity as a Personality Variable," *Journal of Personality* 30(2) (1962).
D. Apter, *Ideology and Discontent* (New York: Free Press, 1964).
E. De St. Aubin, "Personal Ideology Polarity," *Journal of Personality* 71(1) (1996).
H. J. Eysenck and G. D. Wilson, *The Psychological Basis of Ideology* (Lancaster, UK: MTP Press, 1978).
D. L. Hamilton and T. L. Rose, "Illusory Correlation and the Maintenance of Stereotypical Beliefs," *Journal of Personality and Social Psychology* 39(5) (1980).
D. T. Miller and M. Ross, "Self-Serving Biases in the Attribution of Causality: Fact or Fiction?" *Psychological Bulletin* 82(2) (1975).
p. 11 According to Dr. James Watson
Interview with Neil Patterson, president, E. O. Wilson Biodiversity Foundation, Papillon House, Carmel, CA, July 30, 2009.

p. 12 But as social processes, institutions, technologies

Staffora Beer, David Whittaker, and Brian Eno, *Think Before You Think: Social Complexity and Knowledge of Knowing* (Charlbury, UK: Wavestone Press, 2009).

David J. Kinden, *The Accidental Mind: How Brain Evolution Has Given Us Love, Memory, Dreams, and God* (Cambridge, MA: Belknap Press, 2008).

C. M. Hann, *When History Accelerates: Essays on Rapid Social Change and Creativity* (London: Athlone Press, 1994).

Sandra D. Mitchell, *Unsimple Truths: Science, Complexity, and Policy* (Chicago: University of Chicago Press, 2009).

Joseph Tainter, *The Collapse of Complex Societies* (Cambridge: Cambridge University Press, 1990).

Alvin Toffler, *Future Shock* (New York: Bantam Books, 1984).

Glenn M. Schwartz and John J. Nichols, *After Collapse* (Tucson: University of Arizona Press, 2006).

M. Mitchell Waldrop, *Complexity: The Emerging Science at the Edge of Order and Chaos* (New York: Simon and Schuster, 1992).

Melanie Mitchell, *Complexity: A Guided Tour* (New York: Oxford University Press, 2009).

John J. Miller and Scott E. Page, *Complex Adaptive Systems: An Introduction to Computational Models of Social Life* (Princeton, NJ: Princeton University Press, 2007).

Len Fisher, *The Perfect Swarm: The Science of Complexity in Everyday Life* (New York: Basic Books, 2009).

John H. Holland, *Hidden Order: How Adaptation Builds Complexity* (New York: Basic Books, 1996).

Robert Geyer and Samir Rihani, *Complexity and Public Policy: A New Approach to 21st Century Politics, Policy, and Society* (New York: Routledge, 2010).

Linda Elder and Richard Paul, "Critical Thinking in a World of Accelerating Change and Complexity," *Social Education*, December 1, 2008.

Arthur Whimbey and Jack Lochhead, *Problem Solving and Comprehension* (Mahwah, NJ: Lawrence Erlbaum, 1999).

Bryan K. Hanks and Katheryn M. Linduff, *Social Complexity in Prehistoric Eurasia: Monuments, Metals, and Mobility* (New York: Cambridge University Press, 2009).

N. Jausovec and K. Bakracevic, *Creativity Research Journal* 8(1) (January 1995).

G. A. Miller, "The Magical Number Seven, Plus or Minus Two: Some Limits on Our Capacity for Processing Information," *Psychological Review* 63(2) (1956).

p. 14 As the Mayans entered the second phase

Elizabeth Hill Boone and Elizabeth P. Benson, "Ritual Human Sacrifice in Mesoamerica" (Conference at Dumbarton Oaks, Washington, DC, October 13–14, 1979).

David Roberts, "Exploring the Place of Fright," *National Geographic*, July–August 2001.

Vera Tiesler and Andrea Cucina, eds., *New Perspectives on Human Sacrifice and Ritual Body Treatments in Ancient Maya Society* (New York: Springer, 2008).

Matthew Markowitz, "The Mayans, Climate Change, and Conflict," *ICE Case Studies*, No. 112 (2003), http://www1.american.edu/ted/ice/maya.htm.

Charles C. Mann, *1491* (New York: Vintage, 2006).

Glenn Welker, "Mayan Civilization," http://www.indians.org/welker/maya.htm.

Patrick L. Barry, "The Rise and Fall of the Mayan Empire," January 6, 2001, http://

www.firstscience.com/home/articles/origins/the-rise-and-fall-of-the-mayan-empire_
1387.html.

David L. Webster, *The Fall of the Ancient Maya* (New York: Thames and Hudson, 2002).

Michael Lemonick, "Mysteries of the Mayans," http://www.indians.org/welker/maya.htm.

"Collapse: Why Do Civilizations Fail?" http://www.learner.org/interactives/collapse/
mayans.html.

Stefan Lovegren, "Climate Change Killed Off Mayan Civilization, Study Says," *National
Geographic News*, March 13, 2003.

John Ness, "Fall of the Mayan," *Newsweek*, March 24, 2003.

Elin C. Danien and Robert J. Sharer, *New Theories on the Ancient Maya* (Philadelphia:
University Museum, University of Pennsylvania, 1992).

Joseph Tainter, *The Collapse of Complex Societies* (Cambridge: Cambridge University
Press, 1990).

David Freidel and Linda Schele, *A Forest of Kings: The Untold Story of the Ancient Maya*
(New York: HarperPerennial, 1992).

Heather Irene McKillop, *The Ancient Maya: New Perspectives* (New York: Norton, 2006).

Terje Tvedt and Terje Oestigaard, eds., *A History of Water*, Series II, Vol. 1: *Idea of Water
from Ancient Societies to the Modern World* (London: I. B. Tauris, 2009).

Tony Allan and Tom Lowenstein, *Gods of Sun and Sacrifice: Aztec and Maya Myth* (London: Duncan Baird, 1999).

William James, *The Will to Believe* (New York: Book Jungle, 2009).

David Joralemon, *Ritual Blood-Sacrifice Among the Ancient Maya, Part I* (Pebble Beach,
CA: Robert Louis Stevenson School, 1974).

Alex Okeowo, "Portal to Maya Underworld Found in Mexico?" *National Geographic
News*, August 22, 2008.

p. 14 **According to reporter Mark Stevenson**

Mark Stevenson, "Archaeologists Unearth Evidence of Human Sacrifice," Associated
Press, January 22, 2005.

p. 14 **One small Mayan community known as the Lamanai**

Marty Greenberg, Reed Hogan, and Andrew Houshouder, "Mayan Ruins of Belize,"
http://www4.samford.edu/schools/artsci/biology/belize/mayan.html.

David M. Pendergast, *Lamanai Stela 9: The Archaeological Context*, Research Reports on
Ancient Maya Writing 20 (Washington, DC: Center for Maya Research, 1988).

Elizabeth Hill Boone and Elizabeth P. Benson, "Ritual Human Sacrifice in Mesoamerica" (Conference at Dumbarton Oaks, Washington, DC, October 13–14, 1979).

David Roberts, "Exploring the Place of Fright," *National Geographic*, July–August 2001.

Lamanai Archaelogical Project, Indian Church, Belize, http://www.csms.ca/Lamanai%
20Field%20School.htm.

Vera Tiesler and Andrea Cucina, eds., *New Perspectives on Human Sacrifice and Ritual
Body Treatments in Ancient Maya Society* (New York: Springer, 2008).

Matthew Markowitz, "The Mayans, Climate Change, and Conflict," ICE Case Studies,
No. 112 (2003), http://www1.american.edu/ted/ice/maya.htm.

Charles C. Mann, *1491* (New York: Vintage, 2006).

Glenn Welker, "Mayan Civilization," http://www.indians.org/welker/maya.htm.

Patrick L. Barry, "The Rise and Fall of the Mayan Empire," January 6, 2001, http://
www.firstscience.com/home/articles/origins/the-rise-and-fall-of-the-mayan-empire_
1387.html.

David L. Webster, *The Fall of the Ancient Maya* (New York: Thames and Hudson, 2002).

Michael Lemonick, "Mysteries of the Mayans," http://www.indians.org/welker/maya.htm.

"Collapse: Why Do Civilizations Fail?" http://www.learner.org/interactives/collapse/mayans.html.

Stefan Lovegren, "Climate Change Killed Off Mayan Civilization, Study Says," *National Geographic News*, March 13, 2003.

John Ness, "Fall of the Mayan," *Newsweek*, March 24, 2003.

Elin C. Danien and Robert J. Sharer, *New Theories on the Ancient Maya* (Philadelphia: University Museum, University of Pennsylvania, 1992).

Joseph Tainter, *The Collapse of Complex Societies* (Cambridge: Cambridge University Press, 1990).

David Freidel and Linda Schele, *A Forest of Kings: The Untold Story of the Ancient Maya* (New York: HarperPerennial, 1992).

Heather Irene McKillop, *The Ancient Maya: New Perspectives* (New York: Norton, 2006).

Tony Allan and Tom Lowenstein, *Gods of Sun and Sacrifice: Aztec and Maya Myth* (London: Duncan Baird, 1999).

Alex Okeowo, "Portal to Maya Underworld Found in Mexico?" *National Geographic News*, August 22, 2008.

p. 16 Dr. Steven Chu expressed the urgency of the situation

Suzanna Goldenberg, "Obama's Energy Secretary Outlines Dire Climate Change Scenario" *The Guardian*, February 4, 2009.

Norris Hundley, *The Great Thirst* (Berkeley and Los Angeles: University of California Press, 2001).

Mark Reisner, *Cadillac Desert* (New York: Penguin, 1987).

Dorothy Green, *Managing Water* (Berkeley and Los Angeles: University of California Press, 2007).

David Carle, *Introduction to Water in California* (Berkeley and Los Angeles: University of California Press, 2009).

W. Dragoni and B. S. Sukhja, "Climate Change and Groundwater," Geological Society of London Special Publication, No. 288 (May 15, 2008).

Michael Collier and Robert H. Webb, *Floods, Droughts, and Climate Change* (Tucson: University of Arizona Press, 2002).

Steven Solomon, *Water: The Epic Struggle for Wealth, Power, and Civilization* (New York: HarperCollins, 2010).

p. 17 On October 11, 2009, the *New York Times*

"California Lawmakers Again Fail to Reach Water Deal," Associated Press, October 11, 2009.

p. 19 Wilson sees the full picture

Interview with E. O. Wilson, Harvard University, Cambridge, MA, July 1, 2009.

p. 20 Journalist Josh Clark describes one astonishing reaction to immediate threat

Josh Clark, "How Can Adrenaline Help You Lift a 3,500-Pound Car?" http://health.howstuffworks.com/adrenaline-strength.htm.

C. D. Marsden and S. J. C. Meadows, "The Effect of Adrenaline on the Contractions of Human Muscle," *Journal of Physiology* 207(2) (1970).

Ben Martin, "Fight or Flight," February 9, 2006, http://psychcentral.com/lib/2006/fight-or-flight/.

p. 20 Nicholas D. Kristof, a columnist

Nicholas D. Kristof, "When Our Brains Short-Circuit," *New York Times*, July 1, 2009.

p. 21 Dr. Joseph Tainter gives a compelling and realistic description of the fall
Joseph Tainter, *The Collapse of Complex Societies* (Cambridge: Cambridge University Press, 1990).

Benjamin Isaac, *The Invention of Racism in Classical Antiquity* (Princeton, NJ: Princeton University Press, 2004).

Adrian N. Sherwin-White, *Racial Prejudice in Imperial Rome* (Cambridge: Cambridge University Press, 2010).

Adrian Goldsworthy, *How Rome Fell* (New Haven, CT: Yale University Press, 2009).

Donald Kagan, *The End of the Roman Empire: Decline or Transformation?* (Lexington, MA: Heath, 1992).

Bryan Ward-Perkins, *The Fall of Rome* (New York: Oxford University Press, 2005).

G. W. Bowersock, "The Vanishing Paradigm of the Fall of Rome," *Bulletin of the American Academy of Arts and Sciences,* May 1996.

Peter Heather, *The Fall of the Roman Empire* (New York: Oxford University Press, 2006).

p. 23 New archaeological evidence reveals that the Khmer Empire
Richard Stone, "Divining Angkor," *National Geographic,* July 2009.

Claude Jacques and Philippe LaFond, *The Khmer Empire: Cities and Sanctuaries from the 5th to the 13th Century* (Bangkok: River Books, 2007).

Brian M. Fagan, *The Little Ice Age: How Climate Made History* (New York: Basic Books, 2001).

Chapter 2.
Evolution's Gift: A Breakthrough in Neuroscience

p. 27 Jonah Lehrer describes how we break the cognitive threshold
Jonah Lehrer, "The Eureka Hunt," *New Yorker,* July 28, 2008.

Richard C. Rothermel, "Mann Gulch Fire: A Race That Couldn't Be Won," General Technical Report INT-29 (Washington, DC: U.S. Department of Agriculture, Forest Service, May 1993).

Rob Schrepfer and Michael Useem, *Wag Dodge: Leadership Through Innovation or the Evolution of a Failed Leader?* (Chapel Hill, NC: Duke School of Business, Leadership Development Initiative, 2003).

Art Jukkala and Ted Putnam, "Forest Fire Shelters Save Lives," *Fire Management Notes* 47(2) (1986).

Laurence Gonzales, *Deep Survival: Who Lives, Who Dies, and Why* (New York: Norton, 2003).

C. C. Wilson, "Fatal and Near Fatal Forest Fires: The Common Denominators," *International Fire Chief* 43(9) (1977).

Doug Campbell and Bruce Schubert, "The Art of Wildland Firefighting" (Ojai, CA: Working paper, 2009).

p. 28 Then, in 1985, forty years after the Mann Gulch
Richard C. Rothermel, "Mann Gulch Fire: A Race That Couldn't Be Won," General Technical Report INT-29 (Washington, DC: U.S. Department of Agriculture, Forest Service, May 1993).

Richard C. Rothermel and Robert W. Mutch, "Behavior of the Life-Threatening Butte Fire," *Fire Management Notes* 47(2) (1986).

p. 30 Understanding the three methods that modern man uses to solve problems
Mary Lou Decosterd, *Right Brain/Left Brain Leadership: Shifting Style for Maximum Impact* (Westport, CT: Praeger, 2008).

Michael S. Sweeney, *Brain, the Complete Mind: How It Develops, How It Works, and How to Keep It Sharp* (Washington, DC: National Geographic Society, 2009).

Steven Pinker, *The Stuff of Thought* (New York: Penguin, 2007).

Jeff Hawkins, *On Intelligence* (New York: Times Books, 2004).

Rita Carter, *The Human Brain Book* (New York: DK Publishing, 2009).

John Ratey, *A User's Guide to the Brain: Perception, Attention, and the Four Theaters of the Brain* (New York: Vintage, 2002).

Walter J. Freeman, *How Brains Make Up Their Minds* (New York: Columbia University Press, 2001).

M. R. Bennett and P. M. S. Hacker, *The History of Cognitive Neuroscience* (Malden, MA: Wiley-Blackwell, 2008).

Jon Driver, Patrick Haggard, and Tim Shallice, eds., *Mental Processes in the Human Brain* (New York: Oxford University Press, 2008).

Michael Gazzaniga, Richard Ivry, and George R. Mangun, *Cognitive Neuroscience: The Biology of the Mind* (New York: Norton, 2008).

Richard S. J. Frackowiak, Karl J. Friston, Christopher D. Frith, and Raymond J. Dolan, *Human Brain Function* (New York: Academic Press, 1997).

Mark Furman and Fred P. Gallo, *The Neurophysics of Human Behavior: Brain, Mind, Behavior, and Information* (Boca Raton, FL: CRC Press, 2000).

C. J. Shatz, "The Developing Brain," *Scientific American*, September 1992.

B. R. Buchsbaum, S. Greer, W. L. Chang, and K. F. Berman, "Meta-analysis of Neuroimaging Studies of the Wisconsin Card-Sorting Task and Component Processing," *Human Brain Mapping* 25(1) (2005).

Nancy C. Andreasen, *The Creating Brain: The Neuroscience of Genius* (Washington, DC: Dana Press, 2005).

p. 31 In prehistoric times, the ability to detect cheaters

J. S. Winston and B. A. Strange et al., "Automatic and Intentional Brain Responses During Evaluation of Trustworthiness of Faces," *Nature Neuroscience*, February 19, 2002.

James V. Haxby and Elizabeth Hoffman et al., "The Distributed Human Neural System for Face Perception," *Trends in Cognitive Science* 4(6) (2000).

David Livington Smith, *Why We Lie: The Evolutionary Roots of Deception and the Unconscious Mind* (New York: St. Martin's Griffin, 2007).

Interview with Dr. E. O. Wilson, Harvard University, Cambridge, MA, July 1, 2009.

Kim Sterelny, *From Mating to Morality: Evaluating Evolutionary Psychology* (New York: Psychology Press, 2003).

p. 32 After spending years chasing a master forger

Steven Spielberg, director, and Jeff Nathanson, screenplay, *Catch Me If You Can*, Dreamworks Pictures, 2002.

Stan Redding and Frank W. Abagnale, *Catch Me If You Can: The True Story of a Real Fake* (New York: Broadway Books, 2000).

p. 33 Distinguishing an insight from a good idea

Jerry Swartz, "The Conscious 'Pop': A Nonconscious Processing Framework for Problem Solving" (Cold Spring Harbor, NY: Cold Spring Harbor Laboratory, New Frontiers in Studies of the Unconscious, April 9, 2007).

K. J. Gilhooly and P. Murphy, "Differentiating Insight from Non-insight Problems," *Thinking and Reasoning* 11(3) (2005).

Jonah Lehrer, "The Eureka Hunt," *New Yorker*, July 28, 2008.

A. J. K. Pols, "Insight in Problem Solving," http://www.phil.uu.nl/preprints/ckiscripties/SCRIPTIES/018_pols.pdf.

Jing Luo and Guenther Knoblich, "Studying Insight Problem Solving with Neuroscientific Methods," *ScienceDirect* 42(1).

John Kounios and Mark Jung-Beeman, "Brain Activity Differs for Creative and Noncreative Thinkers," *ScienceDaily*, October 29, 2007.

M. Jung-Beeman, E. M. Bowden, J. Haberman, J. L. Frymiare, S. Liu-Arambel et al., "Neural Basis of Solving Problems with Insight," *PLoS*, April 2004.

David Rock and Jeffrey Schwartz, "The Neuroscience of Leadership," *Strategy + Business* 43 (2006).

Jeffrey Schwartz and Sharon Begley, *The Mind and the Brain: Neuroplasticity and the Power of Mental Force* (New York: ReganBooks, 2002).

Jeffrey Schwartz, Henry P. Stapp, and Mario Beauregard, "Quantum Physics in Neuroscience and Psychology: A Neurophysical Model of the Mind-Brain Interaction," *Proceedings of the Royal Society B: Biological Sciences* 360(1458) (June 29, 2005).

Kalina Christoff, Justin M. Ream, and John D. E. Gabrieli, "Neural Basis of Spontaneous Thought Processes," *Cortex* 40(4) (2004).

J. Kounios, J. L. Frymiare, E. M. Bowden, J. I. Fleck, K. Subramaniam, T. B. Parnish, and M. Jung-Beeman, "The Prepared Mind: Neural Activity Prior to Problem Presentation Predicts Subsequent Solution by Sudden Insight," *Psychological Sciences* 17(10) (2006).

A. Dijksterhuis, "A Theory of Unconscious Thought," *Perspectives on Psychological Science* 1(2) (2006).

A. Dijksterhuis, "Think Different: The Merits of Unconscious Thought in Preference Development and Decision Making," *Journal of Personality and Social Psychology* 87(5) (2004).

A. Dijksterhuis, L. Nordgren, and R. Van Baaren, "On Making the Right Choice: The Deliberation-Without-Attention Effect," *Science* 311(5763) (2006).

Jennifer Dorfman, Victor A. Shames, and John F. Kihlstrom, "Intuition, Incubation, and Insight: Implicit Cognition in Problem Solving," in *Implicit Cognition*, ed. Geoffrey Underwood (New York: Oxford University Press, 1996).

Jonathan W. Schooler and Joseph Melcher, "The Ineffability of Insight," in *The Creative Cognition Approach*, ed. Steven M. Smith, Thomas B. Ward, and Ronald A. Fink (Cambridge, MA: MIT Press, 1995).

John Kounios, Jessica I. Fleck, Deborah L. Green, Lisa Payne, Jennifer L. Stevenson, Edward M. Bowden, and Mark Jung-Beeman, "The Origins of Insight in Resting-State Brain Activity" (Philadelphia: Department of Psychology, Drexel University, July 18, 2007).

Yun Chu, *Human Insight Problem Solving: Performance, Processing, and Phenomenology* (Saarbrücken, Germany: VDM Verlag, 2009).

Dianna Amorde and Christine Frank, *Aha! Moments: When Intellect and Intuition Collide* (Boston: Inspired Press, 2009).

Martin Gardner, *Aha! Insight* (New York: Freeman, 1978).

Robert J. Sternberg and Janet E. Davidson, *The Nature of Insight* (Cambridge, MA: MIT Press, 1995).

Robert J. Sternberg and Talia Ben-Zeev, *Complexity Cognition: The Psychology of Human Thought* (New York: Oxford University Press, 2001).

J. G. P. Bargh, A. Lee-Chai, A. Barndollar, and R. Trotschel, "The Automated Will: Nonconscious Activation and Pursuit of Behavioral Goals," *Journal of Personality and Social Psychology* 81(6) (2001).

A. M. Achim, M. C. Bertrand, A. Montoya, A. K. Malla, and M. Lepage, "Medial Temporal Lobe Activation During Associate Memory Encoding for Arbitrary and Semantically Related Objects," *Brain Research* 1161 (2007).

C. Stough, ed., *Neurobiology of Exceptionality* (New York: Kluwer Academic, 2005).

P. I. Ansburg, *Current Psychology* 19(2) (2000), http://www.springerlink.com/content/0flnnd5rjdk3ucy1/.

P. I. Ansburg and R. L. Dominowski, *Journal of Creative Behavior* (2000).

Baker-Sennett and S. J. Ceci, *Journal of Creative Behavior* (1996).

E. M. Bowden and M. J. Beeman, *Psychological Science* (1998).

E. P. Chronicle, Y. C. Omerod, and J. N. MacGregor, *Quarterly Journal of Experimental Psychology Section A—Human Experimental Psychology* 54A(3) (2001).

Robert J. Sternberg and Janet E. Davidson, *The Nature of Insight* (Cambridge, MA: MIT Press, 1995).

A. Kaplan and H. A. Simon, "In Search of Insight," *Cognitive Psychology* 22(3) (1990).

G. Knoblich, S. Ohlsson, H. Haider, and D. Rhenius, *Journal of Experimental Psychology: Learning, Memory, and Cognition*, 1999.

R. S. Lockhart, M. Lamon, and M. L. Gick, *Memory and Cognition*, 1988.

N. R. F. Maier, "Reasoning in Humans II: The Solution of a Problem and Its Appearance in Consciousness," *Journal of Comparative Psychology* 13 (1931).

Richard E. Mayer, *Thinking, Problem Solving, and Cognition* (New York: Freeman, 1992).

J. Metcalfe, *Journal of Experimental Psychology: Learning, Memory, and Cognition* 12(2) (1986).

J. Metcalfe and D. Wiebe, *Memory and Cognition* 15(3) (1987).

N. Mori, *Japanese Psychological Research* (1996).

A. Del Cul, S. Baillet, and S. Dehaene, *PLoS Biology* (2007).

P. Haggard and B. Libel, *Journal of Consciousness Studies* (2001).

Sandkuhler S. Bhattacharya, "Deconstructing Insight: EEG Correlates of Insightful Problem Solving," *PLoS ONE* 3(1) (2008).

Jonathan W. Schooler and Joseph Melcher, "The Ineffability of Insight" (paper presented at the annual meeting of the Psychonomic Society, Washington, DC, 1993).

R. S. Siegler, "Unconscious Insights," *Current Directions in Psychological Science* 9(3) (2000).

R. W. Smith and J. Kounios, "Sudden Insight," *Journal of Experimental Psychology: Learning, Memory, and Cognition* 22(6) (1996).

R. W. Weisberg, *Journal of Experimental Psychology: Learning, Memory, and Cognition* (1992).

M. J. Beeman and E. M. Bowden, *Memory and Cognition* 28(7) (2000).

E. M. Bowden and M. J. Beeman, *Psychological Science* 9(6) (1998).

E. M. Bowden and M. Jung-Beeman, "Aha! Insight Experience Correlates with Solution Activation in the Right Hemisphere," *Psychonomic Bulletin and Review* 10(3) (2003).

E. M. Bowden, M. Jung-Beeman, J. Fleck, and J. Kounios, "New Approaches to Demystifying Insight," *Trends in Cognitive Sciences* 9(7) (2005).

Nancy C. Andreasen, *The Creating Brain: The Neuroscience of Genius* (Washington, DC: Dana Press, 2005).

E. Angelakis, J. F. Lubar, S. Stathopoulou, and J. Kounios, "Peak Alpha Frequency: An Electroencephalographic Measure of Cognitive Preparedness," *Clinical Neurophysiology* 115(4) (2004).

M. Jung-Beeman, E. M. Bowden, J. Haberman, J. L. Frymiare, S. Arambel-Liu, and R.

Greenblatt, "Neural Activity When People Solve Problems with Insight," *PLoS Biology* 2 (2004).

J. Kounios, A. M. Osman, and D. E. Meyer, "Structure and Process in Semantic Memory: New Evidence Based on Speed-Accuracy Decomposition," *Journal of Experimental Psychology* 116(1) (1987).

S. Lang, N. Kanngieger, P. Jaskowski, H. Haider, M. Rose, and R. Verleger, "Precursors of Insight in Event-Related Brain Potentials," *Journal of Cognitive Neuroscience* 18(12) (2006).

A. Newell and H. A. Simon, *Human Problem Solving* (Englewood Cliffs, NJ: Prentice Hall, 1972).

K. Christoff and J. D. E. Gabrieli, "The Frontopolar Cortex and Human Cognition," *Psychobiology* 28(2) (2000).

Geoffrey Underwood, ed., *Implicit Cognition* (New York: Oxford University Press, 1996).

p. 36 To understand the pace at which biological changes

Charles Darwin, *The Descent of Man* (New York: Penguin, 2007).

Charles Darwin, *The Expression of the Emotions in Man and Animals* (New York: Oxford University Press, 1998).

Chris Stringer and Peter Andrews, *The Complete World of Human Evolution* (New York: Thames and Hudson, 2005).

Gregory Cochran and Henry Harpending, *The 10,000 Year Explosion: How Civilization Accelerated Human Evolution* (New York: Basic Books, 2009).

John Cartwright, *Evolution and Human Behavior: Darwinian Perspectives on Human Nature* (Cambridge, MA: MIT Press, 2000).

Eva Jablonka and Marion J. Lamb, *Evolution in Four Dimensions: Genetic, Epigenetic, Behavioral, and Symbolic Variation in the History of Life* (Cambridge, MA: MIT Press, 2005).

Michael J. Behe, *The Edge of Evolution: The Search for the Limits of Darwinism* (New York: Free Press, 2008).

Christopher Scarre, *The Human Past: World Prehistory and the Development of Human Societies* (New York: Thames and Hudson, 2005).

Timothy Goldsmith and William F. Zimmerman, *Biology, Evolution, and Human Nature* (New York: Wiley, 2000).

Philip Clayton and Jeffrey Schloss, *Evolution and Ethics: Human Morality in Biological and Religious Perspective* (Grand Rapids, MI: Eerdmans, 2004).

Bruce H. Lipton and Steve Bhaerman, *Spontaneous Evolution: Our Positive Future (and a Way to Get There from Here)* (Carlsbad, CA: Hay House, 2009).

Richard Restak, *The New Brain: How the Modern Age Is Rewiring Your Mind* (Emmaus, PA: Rodale, 2004).

Mihail C. Roco and Carlo D. Montemagno, *The Coevolution of Human Potential and Converging Technologies* (New York: New York Academy of Science, 2004).

Robert J. Sternberg and Janet E. Davidson, *The Nature of Insight* (Cambridge, MA: MIT Press, 1995).

Harold Jerison, *Evolution of the Brain and Intelligence* (New York: Academic Press, 1973).

p. 37 According to Philip Brownell, renowned biologist

Interview with Dr. Philip Brownell, Oregon State University, Corvallis, OR, May 9, 2009.

p. 38 In support of Brownell, the University of Chicago

"Human Brain Evolution Was a 'Special Event,'" *HHMI Research News*, December 29, 2004, http://www.hhmi.org/news/pdf/lahn3.pdf.

p. 39 Science writer Jonah Lehrer describes the relationship
Jonah Lehrer, "The Eureka Hunt," *New Yorker*, July 28, 2008.

p. 40 Dr. John Kounios, professor of psychology
J. Kounios, J. L. Frymiare, E. M. Bowden, J. I. Fleck, K. Subramaniam, T. B. Parnish, and M. Jung-Beeman, "The Prepared Mind: Neural Activity Prior to Problem Presentation Predicts Subsequent Solution by Sudden Insight," *Psychological Sciences* 17(10) (2006).

John Kounios, Jessica I. Fleck, Deborah L. Green, Lisa Payne, Jennifer L. Stevenson, Edward M. Bowden, and Mark Jung-Beeman, "The Origins of Insight in Resting-State Brain Activity" (Philadelphia: Department of Psychology, Drexel University, July 18, 2007).

Jerry Swartz, "The Conscious 'Pop': A Nonconscious Processing Framework for Problem Solving" (Cold Spring Harbor, NY: Cold Spring Harbor Laboratory, New Frontiers in Studies of the Unconscious, April 9, 2007).

K. J. Gilhooly and P. Murphy, "Differentiating Insight from Non-insight Problems," *Thinking and Reasoning* 11(3) (2005).

Jonah Lehrer, "The Eureka Hunt," *New Yorker*, July 28, 2008.

A. J. K. Pols, "Insight in Problem Solving," http://www.phil.uu.nl/preprints/ckiscripties/SCRIPTIES/018_pols.pdf.

Jing Luo and Guenther Knoblich, "Studying Insight Problem Solving with Neuroscientific Methods," *ScienceDirect* 42(1).

John Kounios and Mark Jung-Beeman, "Brain Activity Differs for Creative and Noncreative Thinkers," *ScienceDaily*, October 29, 2007.

M. Jung-Beeman, E. M. Bowden, J. Haberman, J. L. Frymiare, S. Liu-Arambel et al., "Neural Basis of Solving Problems with Insight," *PLoS* (April 2004).

David Rock and Jeffrey Schwartz, "The Neuroscience of Leadership," *Strategy + Business* 43 (2006).

Jeffrey Schwartz and Sharon Begley, *The Mind and the Brain: Neuroplasticity and the Power of Mental Force* (New York: ReganBooks, 2002).

Jeffrey Schwartz, Henry P. Stapp, and Mario Beauregard, "Quantum Physics in Neuroscience and Psychology: A Neurophysical Model of the Mind-Brain Interaction," *Proceedings of the Royal Society B: Biological Sciences* 360(1458) (June 29, 2005).

Kalina Christoff, Justin M. Ream, and John D. E. Gabrieli, "Neural Basis of Spontaneous Thought Processes," *Cortex* 40(4) (2004).

J. Kounios, J. L. Frymiare, E. M. Bowden, J. I. Fleck, K. Subramaniam, T. B. Parnish, and M. Jung-Beeman, "The Prepared Mind: Neural Activity Prior to Problem Presentation Predicts Subsequent Solution by Sudden Insight," *Psychological Sciences* 17(10) (2006).

A. Dijksterhuis, "A Theory of Unconscious Thought," *Perspectives on Psychological Science* 1(2) (2006).

A. Dijksterhuis, "Think Different: The Merits of Unconscious Thought in Preference Development and Decision Making," *Journal of Personality and Social Psychology* 87(5) (2004).

A. Dijksterhuis, L. Nordgren, and R. Van Baaren, "On Making the Right Choice: The Deliberation-Without-Attention Effect," *Science* 311(5763) (2006).

Jennifer Dorfman, Victor A. Shames, and John F. Kihlstrom, "Intuition, Incubation, and Insight: Implicit Cognition in Problem Solving," in *Implicit Cognition*, ed. Geoffrey Underwood (New York: Oxford University Press, 1996).

Jonathan W. Schooler and Joseph Melcher, "The Ineffability of Insight," in *The Creative Cognition Approach*, ed. Steven M. Smith, Thomas B. Ward, and Ronald A. Fink (Cambridge, MA: MIT Press, 1995).

Yun Chu, *Human Insight Problem Solving: Performance, Processing, and Phenomenology* (Saarbrücken, Germany: VDM Verlag, 2009).

Dianna Amorde and Christine Frank, *Aha! Moments: When Intellect and Intuition Collide* (Boston: Inspired Press, 2009).

Martin Gardner, *Aha! Insight* (New York: Freeman, 1978).

Robert J. Sternberg and Janet E. Davidson, *The Nature of Insight* (Cambridge, MA: MIT Press, 1995).

Robert J. Sternberg and Talia Ben-Zeev, *Complexity Cognition: The Psychology of Human Thought* (New York: Oxford University Press, 2001).

J. G. P. Bargh, A. Lee-Chai, A. Barndollar, and R. Trotschel, "The Automated Will: Nonconscious Activation and Pursuit of Behavioral Goals," *Journal of Personality and Social Psychology* 81(6) (2001).

A. M. Achim, M. C. Bertrand, A. Montoya, A. K. Malla, and M. Lepage, "Medial Temporal Lobe Activation During Associate Memory Encoding for Arbitrary and Semantically Related Objects," *Brain Research* 1161 (2007).

C. Stough, ed., *Neurobiology of Exceptionality* (New York: Kluwer Academic, 2005).

P. I. Ansburg, *Current Psychology* 19(2) (2000).

P. I. Ansburg and R. L. Dominowski, *Journal of Creative Behavior* (2000).

Baker-Sennett and S. J. Ceci, *Journal of Creative Behavior* (1996).

E. M. Bowden and M. J. Beeman, *Psychological Science* (1998).

E. P. Chronicle, Y. C. Omerod, and J. N. MacGregor, *Quarterly Journal of Experimental Psychology Section A—Human Experimental Psychology* 54A(3) (2001).

Knoblich, S. Ohlsson, H. Haider, and D. Rhebius, *Journal of Experimental Psychology: Learning, Memory, and Cognition* (1999).

R. S. Lockhart, M. Lamon, and M. L. Gick, *Memory and Cognition* (1988).

N. R. F. Maier, "Reasoning in Humans II: The Solution of a Problem and Its Appearance in Consciousness," *Journal of Comparative Psychology* 13 (1931).

Richard E. Mayer, *Thinking, Problem Solving, and Cognition* (New York: Freeman, 1992).

J. Metcalfe, *Journal of Experimental Psychology: Learning, Memory, and Cognition* 12(2) (1986).

J. Metcalfe and D. Wiebe, *Memory and Cognition* 15(3) (1987).

N. Mori, *Japanese Psychological Research* (1996).

A. Del Cul, S. Baillet, and S. Dehaene, *PLoS Biology* (2007).

P. Haggard and B. Libel, *Journal of Consciousness Studies* (2001).

Sandkuhler S. Bhattacharya, "Deconstructing Insight: EEG Correlates of Insightful Problem Solving," *PLoS ONE* 3(1) (2008).

Jonathan W. Schooler and Joseph Melcher, "The Ineffability of Insight" (paper presented at the annual meeting of the Psychonomic Society, Washington, DC, 1993).

R. S. Siegler, "Unconscious Insights," *Current Directions in Psychological Science* 9(3) (2000).

R. W. Smith and J. Kounios, "Sudden Insight," *Journal of Experimental Psychology: Learning, Memory, and Cognition* 22(6) (1996).

R. W. Weisberg, *Journal of Experimental Psychology. Learning, Memory, and Cognition* (1992).

M. J. Beeman and E. M. Bowden, *Memory and Cognition* 28(7) (2000).

E. M. Bowden and M. J. Beeman, *Psychological Science* 9(6) (1998).

E. M. Bowden and M. Jung-Beeman, "Aha! Insight Experience Correlates with Solution Activation in the Right Hemisphere," *Psychonomic Bulletin and Review* 10(3) (2003).

E. M. Bowden, M. Jung-Beeman, J. Fleck, and J. Kounios, "New Approaches to Demystifying Insight," *Trends in Cognitive Sciences* 9(7) (2005).

Nancy C. Andreasen, *The Creating Brain: The Neuroscience of Genius* (Washington, DC: Dana Press, 2005).

Angelakis, J. F. Lubar, S. Stathopoulou, and J. Kounios, "Peak Alpha Frequency: An Electroencephalographic Measure of Cognitive Preparedness," *Clinical Neurophysiology* 115(4) (2004).

M. Jung-Beeman, E. M. Bowden, J. Haberman, J. L. Frymiare, S. Arambel-Liu, and R. Greenblatt, "Neural Activity When People Solve Problems with Insight," *PLoS Biology* 2(4) (2004).

J. Kounios, A. M. Osman, and D. E. Meyer, "Structure and Process in Semantic Memory: New Evidence Based on Speed-Accuracy Decomposition," *Journal of Experimental Psychology* 116(1) (1987).

S. Lang, N. Kanngieger, P. Jaskowski, H. Haider, M. Rose, and R. Verleger, "Precursors of Insight in Event-Related Brain Potentials," *Journal of Cognitive Neuroscience* 18(12) (2006).

A. Newell and H. A. Simon, *Human Problem Solving* (Englewood Cliffs, NJ: Prentice Hall, 1972).

K. Christoff and J. D. E. Gabrieli, "The Frontopolar Cortex and Human Cognition," *Psychobiology* 28(2) (2000).

Geoffrey Underwood, ed., *Implicit Cognition* (New York: Oxford University Press, 1996).

p. 41   Despite remarkable strides in solar, wave

"Plans for New Reactor Worldwide," January 2010, http://world-nuclear.org/info/default.aspx?id=416&terms=%E2%80%9CPlans+for+New+Reactor+Worldwide.

U.S. Nuclear Reactors (Washington, DC: Energy Information Administration, 2008).

U.S. DOE budgets figures from Global Energy Network Institute Web site: http://www.geni.org/globalenergy/library/technical-articles/generation/nuclear/scientific american/obama-budget-increases-funding-for-energy-research-and-nuclear-power/index.shtml.

"Obama Budget Increases Funding for Energy Research and Nuclear Power," *Scientific American*, February 2, 2010.

Robert Vandenbosch and Susanne E. Vandenbosch, *Nuclear Waste Stalemate: Political and Scientific Controversies* (Salt Lake City: University of Utah Press, 2007).

Helen Caldicott, *Nuclear Power Is Not the Answer.*(New York: New Press, 2007).

Michio Kaku, *Nuclear Power: Both Sides* (New York: Norton, 1989).

Steven D. Thomas, *The Realities of Nuclear Power: International Economic and Regulatory Experience* (New York: Cambridge University Press, 2010).

Brice Smith, *Insurmountable Risks: The Dangers of Using Nuclear Power to Combat Global Climate Change* (Muskegon, MI: RDR Books, 2006).

p. 42   As of 2010, approximately 15 percent of the power

"Nuclear Power in Canada," April 2010, http://www.world-nuclear.org/info/inf49a_Nuclear_Power_in_Canada.html.

Roger G. Steed, *Nuclear Power in Canada and Beyond* (Renfrew, Ontario: General Store Publishing House, 2007).

p. 42 The same goes for France
"Nuclear Power in France," http://www.world-nuclear.org/info/inf40.html.
Gabrielle Hecht and Michel Callan, *The Radiance of France: Nuclear Power and National Identity After WWII* (Cambridge, MA: MIT Press, 2009).

p. 42 China has also announced plans to build forty nuclear power plants
"China Plans to Build 40 New Nuclear Reactors in Next 15 Years," April 7, 2005, http://www.spacewar.com/2005/050407024531.vn97m1hl.html.

p. 43 Recently, I watched an interview with U.S. Secretary of Energy
Jon Stewart, *The Daily Show*, July 21, 2009.

p. 43 According to Art Rosenfeld, who works on the California Energy Commission
Dana Hull, "Art Rosenfeld, the 'Godfather' of Energy Efficiency," *San Jose Mercury*, December 27, 2009.
Felicity Barringer, "White Roofs Catch on as Energy Cost Cutters," *New York Times*, July 29, 2009.

p. 44 In their 2009 bestselling book *SuperFreakonomics*
Steven D. Levitt and Stephen J. Dubner, *SuperFreakonomics: Global Cooling, Patriotic Prostitutes, and Why Suicide Bombers Should Buy Life Insurance* (New York: HarperCollins, 2009).
Richard Harris, "Scientists Debate Shading Earth as Climate Fix," National Public Radio, June 16, 2009.
Ken Caldeira, *Geoengineering to Shade Earth* (Washington, DC: Worldwatch Institute, 2009).
Jeff Goodell, "Can Dr. Evil Save the World?" *Rolling Stone*, November 16, 2006.

Chapter 3.
The Sovereignty of Supermemes: The Power of Beliefs

p. 47 In a recent interview, Dean Kamen
Isabella Rossellini interview with Dean Kamen, *Iconoclasts*, Sundance Channel, November 16, 2006.
Steve Kemper, *Code Name Ginger: The Story Behind Segway and Dean Kamen's Quest to Invent a New World* (Boston: Harvard Business Press, 2003).
Jim Sidanius and Felicia Pratto, "The Inevitability of Oppression and the Dynamics of Social Dominance," in *Prejudice, Politics, and the American Dilemma*, ed. Paul Sniderman, Philip E. Tetlock, and Edward C. Carmines (Palo Alto, CA: Stanford University Press, 1993).

p. 48 The answer can be found in a little word called a *meme*
Richard Dawkins, *The Selfish Gene* (New York: Oxford University Press, 2006).
Robert Aunger, *The Electric Meme: A New Theory of How We Think* (New York: Free Press, 2002).
Alan Grafen and Mark Ridley, *Richard Dawkins: How a Scientist Changed the Way We Think* (New York: Oxford University Press, 2007).
Susan Blackmore, *The Meme Machine* (New York: Oxford University Press, 1999).
Richard Brodie, *Virus of the Mind: The New Science of the Mind* (Seattle: Integral Press, 1996).
Leigh Hoyle, *Genes, Memes, Culture, and Mental Illness: Toward an Integrative Model* (New York: Springer, 2010).
Aaron Lynch, *Thought Contagion* (New York: Basic Books, 1998).

Kate Distin, *The Selfish Meme: A Critical Reassessment* (New York: Cambridge University Press, 2004).

Robert Aunger, *Darwinizing Culture: The Status of Memetics as Science* (New York: Oxford University Press, 2001).

Stephen Shennan, *Genes, Memes, and Human History: Darwinian Archaeology and Cultural Evolution* (New York: Thames and Hudson, 2003).

H. Keith Henson and Arel Lucas, *Memes, Evolution, and Creationism* (self-published, 1989), http://www.operatingthetan.com/1990-memes.txt.

p. 48 Dr. Susan Blackmore, author of *The Meme Machine*

Susan Blackmore, *The Meme Machine* (New York: Oxford University Press, 1999).

Richard Dawkins, "Foreword" to *The Meme Machine*, by Susan Blackmore (New York: Oxford University Press, 1999).

p. 49 The father of meme theory, Dr. Richard Dawkins

Richard Dawkins, *The Selfish Gene* (New York: Oxford University Press, 1989).

p. 50 In his bestselling book, *Virus of the Mind*, Richard Brodie

Richard Brodie, *Virus of the Mind: The New Science of the Meme* (Seattle: Integral Press, 1996).

p. 51 In the case of the Mayans, once they reached a cognitive threshold

Elizabeth Hill Boone and Elizabeth P. Benson, "Ritual Human Sacrifice in Mesoamerica" (Conference at Dumbarton Oaks, Washington, DC, October 13–14, 1979).

David L. Webster, *The Fall of the Ancient Maya* (New York: Thames and Hudson, 2002).

Michael Lemonick, "Mysteries of the Mayans," http://www.indians.org/welker/maya.htm.

John Ness, "Fall of the Mayan," *Newsweek*, March 24, 2003.

Elin C. Danien and Robert J. Sharer, *New Theories on the Ancient Maya* (Philadelphia: University Museum, University of Pennsylvania, 1992).

Richard Dawkins, *The Selfish Gene* (New York: Oxford University Press, 2006).

Robert Aunger, *The Electric Meme: A New Theory of How We Think* (New York: Free Press, 2002).

Alan Grafen and Mark Ridley, *Richard Dawkins: How a Scientist Changed the Way We Think* (New York: Oxford University Press, 2007).

Susan Blackmore, *The Meme Machine* (New York: Oxford University Press, 1999).

Richard Brodie, *Virus of the Mind: The New Science of the Meme* (Seattle: Integral Press, 1996).

Leigh Hoyle, *Genes, Memes, Culture, and Mental Illness: Toward an Integrative Model* (New York: Springer, 2010).

Aaron Lynch, *Thought Contagion* (New York: Basic Books, 1998).

Kate Distin, *The Selfish Meme: A Critical Reassessment* (New York: Cambridge University Press, 2004).

Robert Aunger, *Darwinizing Culture: The Status of Memetics as Science* (New York: Oxford University Press, 2001).

Stephen Shennan, *Genes, Memes, and Human History: Darwinian Archaeology and Cultural Evolution* (New York: Thames and Hudson, 2003).

David Roberts, "Exploring the Place of Fright," *National Geographic*, July–August 2001.

Vera Tiesler and Andrea Cucina, eds., *New Perspectives on Human Sacrifice and Ritual Body Treatments in Ancient Maya Society* (New York: Springer, 2008).

Matthew Markowitz, "The Mayans, Climate Change, and Conflict," ICE Case Studies, No. 112 (2003), http://www1.american.edu/ted/ice/maya.htm.

Charles C. Mann, *1491* (New York: Vintage, 2006).

Glenn Welker, "Mayan Civilization," http://www.indians.org/welker/maya.htm.

Patrick L. Barry, "The Rise and Fall of the Mayan Empire," January 6, 2001, http://www. firstscience.com/home/articles/origins/the-rise-and-fall-of-the-mayan-empire_1387. html.

"Collapse: Why Do Civilizations Fail?" http://www.learner.org/interactives/collapse/ mayans.html.

Stefan Lovegren, "Climate Change Killed Off Mayan Civilization, Study Says," *National Geographic News*, March 13, 2003.

Joseph Tainter, *The Collapse of Complex Societies* (Cambridge: Cambridge University Press, 1990).

David Freidel and Linda Schele, *A Forest of Kings: The Untold Story of the Ancient Maya* (New York: HarperPerennial, 1992).

Heather Irene McKillop, *The Ancient Maya: New Perspectives* (New York: Norton, 2006).

Charles C. Mann, *1491* (New York: Vintage, 2006).

Terje Tvedt and Terje Oestigaard, eds., *A History of Water*, Series II, Vol. 1: *Idea of Water from Ancient Societies to the Modern World* (London: I. B. Tauris, 2009).

Tony Allan and Tom Lowenstein, *Gods of Sun and Sacrifice: Aztec and Maya Myth* (London: Duncan Baird, 1999).

p. 51 Complexity also led to pervasive beliefs

Walker Wakefield and Austin Evans, *Heresies of the High Middle Ages* (New York: Columbia University Press, 1991).

Andre Vauchez, *The Laity in the Middle Ages: Religious Beliefs and Devotional Practices* (South Bend, IN: University of Notre Dame Press, 1996).

Regine Pernoud and Anne Englund Nash, *Those Terrible Middle Ages: Debunking the Myths* (San Francisco: Ignatius Press, 2000).

Richard Gameson and Henriett Leyser, *Belief and Culture in the Middle Ages*, Studies Presented to Henry Mayr–Harting (New York: Oxford University Press, June 14, 2001).

Edward Peters, *Heresy and Authority in Medieval Europe* (Philadelphia: University of Pennsylvania Press, 1980).

Alister E. McGrath, *Heresy: A History of Defending the Truth* (New York: HarperOne, 2009).

John H. Holland, *Hidden Order: How Adaptation Builds Complexity* (New York: Basic Books, 1996).

Staffora Beer, David Whittaker, and Brian Eno, *Think Before You Think: Social Complexity and Knowledge of Knowing* (Charlbury, UK: Wavestone Press, 2009).

David J. Kinden, *The Accidental Mind: How Brain Evolution Has Given Us Love, Memory, Dreams, and God* (Cambridge, MA: Belknap Press, 2008).

C. M. Hann, *When History Accelerates: Essays on Rapid Social Change and Creativity* (London: Athlone Press, 1994).

Sandra D. Mitchell, *Unsimple Truths: Science, Complexity, and Policy* (Chicago: University of Chicago Press, 2009).

Joseph Tainter, *The Collapse of Complex Societies* (Cambridge: Cambridge University Press, 1990).

Richard Dawkins, *The Selfish Gene* (New York: Oxford University Press, 2006).

Robert Aunger, *The Electric Meme: A New Theory of How We Think* (New York: Free Press, 2002).

Alan Grafen and Mark Ridley, *Richard Dawkins: How a Scientist Changed the Way We Think* (New York: Oxford University Press, 2007).

Susan Blackmore, *The Meme Machine* (New York: Oxford University Press, 1999).

Richard Brodie, *Virus of the Mind: The New Science of the Meme* (Seattle: Integral Press, 1996).

Leigh Hoyle, *Genes, Memes, Culture, and Mental Illness: Toward an Integrative Model* (New York: Springer, 2010).

Aaron Lynch, *Thought Contagion* (New York: Basic Books, 1998).

Kate Distin, *The Selfish Meme: A Critical Reassessment* (New York: Cambridge University Press, 2004).

Robert Aunger, *Darwinizing Culture: The Status of Memetics as Science* (New York: Oxford University Press, 2001).

Stephen Shennan, *Genes, Memes, and Human History: Darwinian Archaeology and Cultural Evolution* (New York: Thames and Hudson, 2003).

p. 52 I was introduced to David J. Leinweber

David J. Leinweber, *Nerds on Wall Street: Math, Machines, and Wired Markets* (Hoboken, NJ: Wiley, 2009).

p. 55 John Spence Weir, who has been visually documenting the modern history of Mexico

Interview with John Spence Weir, San Ramon, CA, 1997.

p. 57 The fragile Smith's Blue had become dependent on a single food source

Interview with Richard A. Arnold, Carmel, CA, 2003.

Interview with biologist Jeffrey Norman, Carmel, CA, 2003–2004.

M. Lambert, Miriam Surhone, T. Timpledon, and Susan F. Marseken, *Smith's Blue Butterfly, Maritime Coast Range Ponderosa Pine Forests, Monterey Bay, Overpopulation, Highway, Carnonera Creek, Invasive Plant* (Beau Bassin, Mauritius: Betascript, 2010).

Richard A. Arnold, "Low-Effect Habitat Conservation Plan for the Smith's Blue Butterfly, Wildcat Line Property, Carmel Highlands, Monterey County, California, March 1999," http://www.fws.gov/ventura/endangered/hconservation/hcp/hcfiles/Sarment/Low%20effect%20HCP%20for%20Sarment%20Parcel.pdf.

Zander Associates, "Biological Resources of the Del Monte Forest: Special-Status Species" (Report prepared for Pebble Beach Company, July 2001).

Harold A. Mooney, J. Hall Cushman, Ernesto Medina, Osvaldo Esala, and Ernest-Detlef Schultz, *Functional Role of Biodiversity: A Global Perspective* (New York: Wiley, 1996).

Joseph Zbilut, *Unsustainable Singularities and Randomness: Their Importance in the Complexity of Physical, Biological, and Social Sciences*, http://science.niuz.biz/ebooks-t136212.html?s=5fdee0fd1c3ea106d63f9c496a79215d&amp.

Ray Kurzwell, *The Singularity Is Near: When Humans Transcend Biology* (New York: Penguin, 2006).

p. 58 Dr. Yaneer Bar-Yam of Harvard University

Yaneer Bar-Yam, *Making Things Work* (Cambridge, MA: Knowledge Press, 2004).

p. 59 The real obstacles are our "attitudes"

Isabella Rossellini interview with Dean Kamen, *Iconoclasts*, Sundance Channel, November 16, 2006.

p. 59 Remember the case of criminal mastermind

Steven Spielberg, director, and Jeff Nathanson, screenplay, *Catch Me If You Can*, Dreamworks Pictures, 2002.

Stan Redding and Frank W. Abagnale, *Catch Me If You Can: The True Story of a Real Fake* (New York: Broadway Books, 2000).

p. 60 How similar the economies of many industrialized countries have become

Matthew Franklin, "G20 Leaders Call for Global Economics Enforcer," *The Australian*, April 3, 2009.

Serge Latouche, *The Westernization of the World: The Significance, Scope, and Limits of the Drive Towards Global Uniformity* (Cambridge: Polity Press, 1996).

Rudolf Steiner and Christopher Houghton Budd, *Economics: The World as One Economy* (Canterbury, UK: New Economy, 1996).

"IFRS Promotes Greater Transparency and Uniformity," *The Star*, April 20, 2009.

William Greider, *One World Ready or Not: The Manic Logic of Global Capitalism* (New York: Touchstone,1998).

Mark B. Smith, *History of the Global Stock Market: From Ancient Rome to Silicon Valley* (Chicago: University of Chicago Press, 2003).

George Akerlof and Robert J. Shiller, *Animal Spirits: How Human Psychology Drives the Economy and Why It Matters or Global Capitalism* (Princeton, NJ: Princeton University Press, 2010).

p. 61 Take global stock markets

"Fear Grips Global Stock Markets," BBC, October 10, 2008.

"Global Stock Markets Plunge," *World Watch*, October 2007.

Pan Pylas, "World Stock Markets Tumble amid Fears of Slow Global Growth," *Daily Star*, January 27, 2010.

Kevin Sullivan and Edward Cody, "World's Stock Market Plunge," *Washington Post*, October 7, 2008.

"Stock Markets Plunge Worldwide," CBS News, January 21, 2008.

Melvin Porter, *Financial Crises: A Detailed View on Financial Crises Between 1929 and 2009* (Scotts Valley, CA: CreateSpace, 2009).

"Can the World Stop the Slide?" *Time*, February 4, 2008.

p. 64 Author Richard Brodie acknowledges the power of awareness

Richard Brodie, *Virus of the Mind: The New Science of the Meme* (Seattle: Integral Press, 1996).

Chapter 4.
Irrational Opposition: The First Supermeme

p. 68 This was no longer just a handful of students

Michelle Garcia, "Thousands in Manhattan Protest War," *Washington Post*, March 21, 2004.

p. 71 Prison overcrowding had produced such hazardous health

Jenifer Warren, "CA: Prisoners Sue to Limit State Prison Populations," *Los Angeles Times*, November 16, 2006.

Don Thompson, "Judges Tentatively Order California Inmate Release," Associated Press, February 9, 2009.

Carol F. Williams, "Federal Judges Order California to Release 43,000 Inmates," *Los Angeles Times*, August 5, 2009.

Arnold Schwarzenegger, "Overcrowding State of Emergency Proclamation" (Sacramento: Office of the Governor, October 4, 2006).

Jennifer Steinhauer, "California to Address Prison Overcrowding with Giant Building Program," *New York Times*, April 27, 2007.

Alison Stateman, "California's Prison Crisis: Be Very Afraid," *Time*, August 14, 2009.

Bert Useemand and Anne Morrison Piehl, *Prison State: The Challenge of Mass Incarceration* (New York: Cambridge University Press, 2008).

Mike Hough, Rob Allen, and Enver Solomon, *Tackling Prison Overcrowding: Build More Prisons? Sentence Fewer Offenders?* (Bristol, UK: Policy Press, 2008).

James Austin, *The Growing Imprisonment of California* (Oakland, CA: National Council on Crime and Delinquency, 1986).

Franklin E. Zimring and Gordon Hawkins, *Prison Population and Criminal Justice in California* (Berkeley: Institute of Governmental Studies Press, University of California, 1992).

"The Problem of Prison Overcrowding and Its Impact on the Criminal Justice System" (hearing before the Subcommittee on Penitentiaries and Corrections, Congress, 95th Cong., 1st sess., December 13, 1977).

p. 71 It didn't take long, however, for the state's decision

Mike Rhodes, "What Have Prisons Done to the Valley?" (people's hearing in Fresno, California, February 27, 2005), www.paglen.org/carceral/pdfs/craig_gilmore.pdf.

Rachael McGrath, "Group Rallies Opposition to Proposed Camarillo Prison Site," *Ventura County Star*, June 8, 2008.

Evelyn Nieves, "Storm Raised by Plan for a California Prison," *New York Times*, August 27, 2000.

p. 73 According to the Department of Justice

U.S. Department of Justice, "Criminal Offenders Statistics" (Washington, DC: U.S. Department of Justice, October 2009).

p. 74 Today less than 3 percent of the prison budget

Figures derived budget numbers on the Bureau of Justice Statistics Web site: http://www.ojp.usdoj.gov/bjs/pub/ascii/spe96.txt.

p. 75 Politicians, for example, are masters at using this oppositional approach

Hans J. Eysenck, *The Psychology of Politics* (London: Routledge and Kegan Paul, 1954).

Michael A. Milburn, *Persuasion and Politics: The Social Psychology of Public Opinion* (Pacific Grove, CA: Brooks-Cole/Wadsworth, 1991).

Jeffrey M. Jones, "Obama Approval Continues to Show Party, Age, Race Gaps" Gallup Web site, May 11, 2010, http://www.gallup.com/poll/127481/obama-approval-continues-show-party-age-race-gaps.aspx.

p. 80 The same goes for adults

David Rock and Jeffrey Schwartz, "The Neuroscience of Leadership," *Strategy + Business* 43 (2006).

Charles K. Ogden, *Opposition: A Linguistic and Psychological Analysis* (Bloomington: Indiana University Press, 1967).

p. 81 Inside every factory there are routine tasks

Michael S. Sweeney, *Brain, the Complete Mind: How It Develops, How It Works, and How to Keep It Sharp* (Washington, DC: National Geographic Society, 2009).

Steven Pinker, *The Stuff of Thought* (New York: Penguin, 2007).

Jeff Hawkins, *On Intelligence* (New York: Times Books, 2004).

Rita Carter, *The Human Brain Book* (New York: DK Publishing, 2009).

John Ratey, *A User's Guide to the Brain: Perception, Attention, and the Four Theaters of the Brain* (New York: Vintage, 2002).

Walter J. Freeman, *How Brains Make Up Their Minds* (New York: Columbia University Press, 2001).

M. R. Bennett and P. M. S. Hacker, *The History of Cognitive Neuroscience* (Malden, MA: Wiley-Blackwell, 2008).

Jon Driver, Patrick Haggard, and Tim Shallice, eds., *Mental Processes in the Human Brain* (New York: Oxford University Press, 2008).

Michael Gazzaniga, Richard Ivry, and George R. Mangun, *Cognitive Neuroscience: The Biology of the Mind* (New York: Norton, 2008).

Richard S. J. Frackowiak, Karl J. Friston, Christopher D. Frith, and Raymond J. Dolan, *Human Brain Function* (New York: Academic Press, 1997).

Patrick McNamara, *The Neuroscience of Religious Experience* (New York: Cambridge University Press, 2009).

Mary Lou Decosterd, *Right Brain/Left Brain Leadership: Shifting Style for Maximum Impact* (Westport, CT: Praeger, 2008).

Michael S. Sweeney, *Brain, the Complete Mind: How It Develops, How It Works, and How to Keep It Sharp* (Washington, DC: National Geographic Society, 2009).

Mark Furman and Fred P. Gallo, *The Neurophysics of Human Behavior: Brain, Mind, Behavior, and Information* (Boca Raton, FL: CRC Press, 2000).

p. 82 In a 2006 article, "The Neuroscience of Leadership"

David Rock and Jeffrey Schwartz, "The Neuroscience of Leadership," *Strategy + Business* 43 (2006).

p. 84 Beginning in 1938 with Pavlov's early experiments

Daniel Philip Todes, *Pavlov, Ivan: Exploring the Animal Machine* (New York: Oxford University Press, 2000).

B. F. Skinner, *About Behaviorism* (New York: Vintage, 1974).

B. F. Skinner, *Beyond Freedom and Dignity* (Indianapolis, IN: Hackett,1971).

Kerry W. Buckley, *Mechanical Man: John B. Watson and the Beginning of Behaviorism* (New York: Guilford Press, 1984).

Paul Naour, E. O. Wilson, and B. F. Skinner, *A Dialogue Between Sociobiology and Radical Behaviorism* (New York: Springer, 2009).

p. 85 In the words of chemist and author Linus Pauling

Quoteland.com, November 2009.

Chapter 5.
The Personalization of Blame: The Second Supermeme

p. 87 At the end of 2009 a twenty-three-year-old Nigerian

"Suspect Charged in Airline Bombing Attempt," CBS/Associated Press, December 26, 2009.

Anahad O'Connor and Eric Schmidt, "U.S. Says Passenger Tried to Detonate Device," *New York Times*, December 26, 2009.

p. 88 He went looking for individual culprits

Peter Baker and Carl Hulse, "U.S. Had Early Signs of a Terror Plot, Obama Says," *New York Times*, December 29, 2009.

Susan Davis, "A Guide to the Blame Game Following Failed Terror Attack," *Washington Wire*, December 30, 2009.

Debbie Schlussel, "Why Is Hillary Clinton Escaping Flight 253 Blame?" December 30, 2009, www.debbieschlussel.com/15050/hey-maybe-one-day-shell-come-up-with-her-own-ideas/.

Jeff Zeleny and Helen Cooper, "Obama Says Plot Could Have Been Disrupted," *New York Times*, January 5, 2010.

Peter Baker, "Obama Cites 'Systemic Failure' in U.S. Security," *New York Times*, December 29, 2009.

Bartholomew Elias, *Airport and Aviation Security: U.S. Policy and Strategy in the Age of Global Terrorism* (Boca Raton, FL: Auerbach, 2009).

Kathleen M. Sweet, *Aviation and Airport Security: Terrorism and Safety Concerns* (Boca Raton, FL: CRC Press, 2008).

J. E. Dittes, "Impulsive Closure as a Reaction to Failure-Induced Threat," *Journal of Abnormal and Social Psychology* 63(3) (1961).

p. 91 In 2008, on the verge of bankruptcy

John D. Stoll, Matthew Dolan, Jeffrey McCracken, Josh Mitchell, "Big Three Seek $34 Billion Aid," *Wall Street Journal*, December 3, 2008.

Brian Ross and Joseph Rhee, "Big Three CEOs Flew Private Jets to Plead for Public Funds," ABC News, November 19, 2008.

Dana Mibank, "Auto Execs Fly Corporate Jets to D.C., Tin Cups in Hand," *Washington Post*, November 20, 2008.

Micheline Maynard, *End of Detroit: How the Big Three Lost Their Grip on the American Car Market* (New York: Broadway Business, 2004).

p. 92 The CEO of General Motors, Rick Wagoner, became the predictable scapegoat

Ken Besinger and Jim Buzzanghera, "General Motors CEO Rick Wagoner to Step Down," *Los Angeles Times*, March 30, 2009.

"Obama Fires GM's CEO," Associated Press, March 29, 2009.

Paul Tharp and Andy Geller, "Obama 'Fires' GM Boss," *New York Post*, March 30, 2009.

J. E. Dittes, "Impulsive Closure as a Reaction to Failure-Induced Threat," *Journal of Abnormal and Social Psychology* 63(3) (1961).

p. 93 The nation had a similar reaction

Karen Frefield, "AIG Gives Connecticut's Blumenthal Data on Bonuses," March 21, 2009, www.bloomberg.com/apps/news?pid=20601087&sid=aqoxOvvxmRMU.

David Cho and Brady Dennis, "Bailout King AIG Still to Pay Millions in Bonuses," *Washington Post*, March 15, 2009.

Ronald Shelp and Al Ehrbar, *Fallen Giant: The Amazing Story of Hank Greenberg and the History of AIG* (Hoboken, NJ: Wiley, 2009).

Andrew Spencer, *Tower of Thieves: Inside AIG's Culture of Corporate Greed* (New York: Brick Tower Books, 2009).

American International Group, Inc., *Background Report*, www.choicelevel.com.

J. E. Dittes, "Impulsive Closure as a Reaction to Failure-Induced Threat," *Journal of Abnormal and Social Psychology* 63(3) (1961).

p. 93 The CEO had wisely accepted his post at AIG

Kenneth Musante, "AIG Chief Slashes Salary to $1," CNNMoney, November 25, 2008.

p. 94 According to Anne Szustek

Anne Szustek, "AIG's Federal Bailout Funds Now Total $152.5 Billion," Associated Press, November 10, 2008.

Jonathan Weisman, Sudeep Reddy, Liam Pleven, "Political Heat Sears AIG," *Wall Street Journal*, March 17, 2009.

p. 96 Not surprisingly, recent data released by Marketdata

John LaRosa, "The U.S. Market for Self-Improvement Products and Services," www.prwebdirect.com/releases/2006/9/prweb440011.htm.

p. 98 In an articled titled "Using Willpower to Your Advantage"

Allie Firestone, "Eyes on the Prize: Using Willpower to Your Advantage," divinecaroline.com, http://partners.realgirlsmedia.com/22189/90383-eyes-prize-using-willpower-advantage/3.

p. 99 Less than 3 percent of all garbage generated by Americans is municipal waste
"Waste Recycling: Data, Maps, and Graphs," http://space-age-recycle-solutions.com/facts_pages.htm.

R. W. Beck, "U.S. Recycling Economic Information Study" (Washington, DC: National Recycling Coalition, July 2001).

Mitchell Young, *Garbage and Recycling: Opposing Viewpoints* (San Diego, CA: Greenhaven Press, 2007).

Jennifer Carless, *Taking Out the Trash: A No-Nonsense Guide to Recycling* (Washington, DC: Island Press, 1992).

Frank Aakerman, *Why Do We Recycle? Markets, Values, and Public Policy* (Washington, DC: Island Press, 1996).

p. 99 The facts don't support our paranoia
Robert D. Hormars, Testimony of Robert D. Hormars, Vice Chairman of Goldman Sachs (International) before Committee on Finance, U.S. Senate, March 29, 2006.

Steven Kamin, Mario Marazzi, and John W. Schindler, "Is China 'Exporting Deflation'?" *Federal Reserve Board*, no. 791 (January 2004).

Derek Scissors, "U.S.-China Trade: Do's and Don'ts for Congress" (Washington, DC: Heritage Foundation, July 20, 2009).

Craig K. Elwell, Marc Labonte, and Wayne M. Morrison, "Is China a Threat to U.S. Economy?" (Washington, DC: Congressional Research Service Report for Congress, January 23, 2007).

p. 102 Oprah took a courageous stance on obesity
Lee Ferran, "O's Heavy-Skinny Oprah Cover Shows Weight Contrast," ABC News Online, December 10, 2008, http://abcnews.go.com/GMA/Diet/story?id=6431388&page=1.

Oprah Winfrey, "How Did I Let This Happen Again?" *O, the Oprah Magazine*, January 2009.

p. 103 According to the Centers for Disease Control
Holly L. Roberts, "Obesity in U.S. Children," www.livestrong.com/article/103926-cause-obesity-children/.

CDC, "Obesity Prevalence 25 Percent or Higher in 32 States" (Washington, DC: Centers for Disease Control and Prevention, July 8, 2009).

Morgan Spurlock, *Super Size Me*, distribution by Samuel Goldwyn, Roadside Attractions, May 7, 2004.

Peter Kopelman, Ian Caterson, and William Dietz, *Clinical Obesity in Adults and Children* (Hoboken, NJ: Wiley-Blackwell, 2009).

Lisa Tartamella, Elaine Herscher, and Chris Woolston, *Generation Extra Large: Rescuing Our Children from the Epidemic of Obesity* (New York: Basic Books, 2006).

p. 103 In 1997, the World Health Organization
Theresa Maher, "America's Obesity Crisis: Any Solutions?" *News Locale*, August 29, 2007.

p. 104 During prehistoric times, the acquisition and judicious use of calories was vital
Stephen Mennell, Ann Murcott, and Anneke van Otterloo, *The Sociology of Food: Eating, Diet, and Culture* (Thousand Oaks, CA: Sage, 1992).

Interview with Dr. E. O. Wilson, Harvard University, Cambridge, MA, July 1, 2009.

John Ratey, presentation at Renaissance Conference, Monterey, CA, September 4, 2009.

Interview with John Ratey, Cambridge, MA, October 8, 2009.

Charles Darwin, *The Descent of Man* (New York: Penguin, 2007).

Charles Darwin, *The Expression of the Emotions in Man and Animals* (New York: Oxford University Press, 1998).

Chris Stringer and Peter Andrews, *The Complete World of Human Evolution* (New York: Thames and Hudson, 2005).

Gregory Cochran and Henry Harpending, *The 10,000 Year Explosion: How Civilization Accelerated Human Evolution* (New York: Basic Books, 2009).

John Cartwright, *Evolution and Human Behavior: Darwinian Perspectives on Human Nature* (Cambridge, MA: Bradford/Macmillan, 2000).

Michael J. Behe, *The Edge of Evolution: The Search for the Limits of Darwinism* (New York: Free Press, 2008).

Timothy Goldsmith and William F. Zimmerman, *Biology, Evolution, and Human Nature* (New York: Wiley, 2000).

Nicholas Wade, "Researchers Say Human Brain Is Still Evolving," *New York Times*, September 8, 2005.

William H. Calvin, Terrence Deacon, Ralph L. Holloway, Richard G. Klein, Steven Pinker, John Tooby, Endel Tulving, and Ajit Varki, "The Evolution of the Human Brain" (Center for Human Evolution proceedings of Workshop 5, Bellevue, WA, March 19–20, 2005), www.futurefoundation.org/documents/che_pro_wrk5.pdf.

"Human Brain Evolution Was a 'Special Event,'" *HHMI Research News*, December 29, 2004, http://www.hhmi.org/news/pdf/lahn3.pdf.

Jane Bradbury, "Molecular Insights into Human Brain Evolution," *PLoS Biology* 3(3) (2005).

Kate Melville, "Evolution of the Human Brain Unique," Scienceagogo, December 29, 2004, http://www.scienceagogo.com/news/20041129182724data_trunc_sys.shtml.

George F. Striedter, *Principles of Brain Evolution* (Sunderland, MA: Sinauer Associates, 2004).

Christopher Willis, *The Runaway Brain: The Evolution of Human Uniqueness* (New York: Basic Books, 1994).

p. 105 Ratey explained the details of a recent study

John Ratey and Eric Hagerman, *Spark: The Revolutionary New Science of Exercise and the Brain* (New York: Little, Brown, 2008).

Frawley Bridwell, "Study Shows Link Between Morbid Obesity, Low IQ in Toddlers" (Gainesville: College of Medicine, University of Florida, August 2008).

Paul Thompson, Cyrus A. Raji, "Brain Structure and Obesity," *Human Brain Mapping* 31(3) (March 2010).

Theresa Maher, "America's Obesity Crisis: Any Solutions?" *News Locale*, August 29, 2007.

p. 107 When B. F. Skinner published *Beyond Freedom and Dignity*

B. F. Skinner, *Beyond Freedom and Dignity* (Indianapolis, IN: Hackett,1971).

B. F. Skinner, *About Behaviorism* (New York: Vintage,1974).

Daniel Philip Todes, *Pavlov, Ivan: Exploring the Animal Machine* (New York: Oxford University Press, 2000).

Kerry W. Buckley, *Mechanical Man: John B. Watson and the Beginning of Behaviorism* (New York: Guilford Press, 1984).

Paul Naour, E. O. Wilson, and B. F. Skinner, *A Dialogue Between Sociobiology and Radical Behaviorism* (New York: Springer, 2009).

p. 109 Business consultant David Gurteen, who calls himself a "Knowledge Management Facilitator"
David Gurteen, Hampshire, United Kingdom, 2009.

Chapter 6.
Counterfeit Correlation: The Third Supermeme

p. 111 The Japanese eat very little fat
"Medical Humor: Heart Disease Explained," e-jokes.net.

p. 112 The third supermeme and obstacle to progress is *counterfeit correlation*
James Burrows, Glen Charles, and Les Charles, *Cheers*, Paramount Television for NBC, 1982–1993.

p. 114 A few years ago, journalist Charlene Laino made the following report
Charlene Laino, "Cell Phones Disrupt Teens' Sleep," CBS News, June 9, 2008.

p. 116 At one time we were convinced Hormone Replacement Therapy
Michael Marsh, Malcolm Whitehead, and John Stevenson, *HRT and Cardiovascular Disease* (London: Martin Dunitz, 1996).

P. Collins and C. M. Beale, *The Cardioprotective Role of HRT: A Clinical Update* (New York: Informa Healthcare, November 2004).

Mayer Eisenstein, *Unavoidably Dangerous: Medical Hazards of Synthesized HRT* (CMI Press, 2002).

S. Chew and S. C. Ng, "Hormone Replacement Therapy (HRT) and Ischemic Heart Disease," *Singapore Medical Journal* 164(9) (2002).

Andrea Genazzani, *Hormone Replacement Therapy and Cardiovascular Disease: The Current Status of Research and Practice* (Pearl River, NY: Parthenon, 2001).

Elizabeth Siegel Watkins, *The Estrogen Elixir: A History of Hormone Replacement Therapy in America* (Baltimore, MD: Johns Hopkins University Press, 2007).

CNW Research: http://cnwmr.com/nss-folder/automotiveenergy.

Matt Power, "Don't Buy that New Prius..." *Wired Magazine*, May 19, 2008.

p. 116 Recently I came across a Web site maintained by Dr. Jon Mueller
Jonathan Mueller, "Headlines of Public Press Articles," jonathan.mueller.faculty.noctrl.edu, http://jonathan.mueller.faculty.noctrl.edu/100/correlation_or_causation.htm.

p. 118 In 2002, California reported having almost thirty-five million residents
U.S. Census Bureau, "State and County Quick Facts: California Population" (Washington, DC: U.S. Census Bureau, 2007).

p. 120 New standardized testing shows a dangerous decline
Thomas Toch, *In the Name of Excellence: The Struggle to Reform the Nation's Schools, Why It's Failing, and What Should Be Done* (New York: Oxford University Press, 1991).

Jerry Wartgow, *Why School Reform Is Failing and What We Need to Do About It: 10 Lessons from the Trenches* (Lanham, MD: Rowman and Littlefield Education, 2007).

Emily Forrest Cataldi, Jennifer Laird, Angelina Kewal Ramani, and Chris Chapman, *High School Dropout and Completion Rates in the United States: 2007* (Washington, DC: United States Department of Education, 2007).

Elaine K. McEwan, *Angry Parents, Failing Schools: What's Wrong with the Public Schools and What You Can Do About It* (Tunbridge Wells, UK: Shaw Books, 2000).

Phillip Kaufman, Jin Y. Kwon, Steve Klein, MRP Associates, and Christopher Chapman, *Dropout Rates in the United States, 2000* (Washington, DC: National Center for Education Statistics, 2000).

Samuel L. Blumenfeld, *The Whole Language/OBE Fraud: The Shocking Story of How America Is Being Dumbed Down by Its Own Education System* (Boulder, CO: Paradigm Publishers, 1995).

Robert J. Manley, *Designing School Systems for All Students: A Tool Box to Fix America's Schools* (Lanham, MD: Rowman and Littlefield Education, 2009).

Hank Kraychir, *Destruction of the Education Monster: The Destruction of America's Public School System* (CreateSpace, 2008).

p. 120 One of the unhealthy consequences of the high dropout rate

Stanley Kurtz, "Deflating Grade Inflation," *National Review*, September 27, 2006.

"National Trends in Grade Inflation: American Colleges and Universities," March 14, 2009, www.gradeinflation.com.

Valene Johnson, *Grade Inflation: Crisis in College Education* (New York: Springer, 2003).

Lester H. Hunt, *Grade Inflation: Academic Standards in Higher Education* (Albany: State University of New York Press, 2009).

p. 122 Dr. Yaneer Bar-Yam summarizes the dangerous role false correlations play

Yaneer Bar-Yam, *Making Things Work* (Cambridge, MA: Knowledge Press, 2004).

p. 123 Growing number of children who are now depressed

Mary H. Sarafolean, PhD, "Depression in School-Age Children and Adolescents: Characteristics, Assessment and Prevention," *HealthyPlace.com*, January 2, 2009.

"Teen Suicide Rate: Highest in 15 Years," *Science Daily*, September 8, 2007.

Kevin Johnson, "Cities Grapple with Crimes by Kids," *USA Today*, July 12, 2006.

p. 123 According to Parry, insufficient mitigations lead to "adaptation deficits"

Kimery Wiltshire, "Climate Change of Western Water: Field Analysis Synopsis," October, 1, 2008, www.exloco.org/projects/Carpe_Diem_Oct_2008_Situation_Overview.pdf.

Akima Sumi, Kensuke Fukushi, and Ai Hiramatsu, eds., *Adaptation and Mitigation Strategies for Climate Change* (Tokyo: Springer, 2010).

Adam Zachary Rose, *The Economics of Climate Change Policy: International, National, and Regional Mitigation Strategies* (Cheltenham, UK: Edward Elgar, 2009).

Andrew Jordan, Dave Huitema, Harro Van Asselt, and Tim Rayner, *Climate Change Policy in the European Union: Confronting the Dilemmas of Mitigation and Adaptation?* (Cambridge: Cambridge University Press, 2010).

Roy W. Spencer, *Climate Confusion: How Global Warming Hysteria Leads to Bad Science, Pandering, and Misguided Policies That Hurt the Poor* (New York: Encounter Books, 2010).

p. 126 When reverse engineering is applied to highly complex problems

A. J. McEvily, *Reverse Engineering Gone Wrong: A Case Study* (Maryland Heights, MO: Elsevier, March 4, 2005).

Frederic P. Miller, Agnes F. Vandome, and John McBrewster, *False Dilemma: Fallacy, Dichotomy, Wishful Thinking, Collectively Exhaustive Events, Mutually Exclusive Events, Fuzzy Logic, Principle of Ambivalence, Correlative-Based Fallacies, Degree of Truth* (Beau Bassin, Mauritius: Alphascript, 2009).

p. 127 According to the founders of Wikipedia

www.wikipedia.com.

http://stats.wikimedia.org/reportcard/.

Frederic Miller, Agnes F. Vandome, and John McBrewster, *Criticism of Wikipedia* (Beau Bassin, Mauritius: Alphascript, 2009).

Andrew Lih, *The Wikipedia Revolution: How a Bunch of Nobodies Created the World's Greatest Encyclopedia* (New York: Hyperion, 2009).

p. 128 I had the opportunity to watch the former Chairman of the Federal Reserve

David Weir, "Fact-Checking Alan Greenspan," October 26, 2007, http://hotweir.blogspot.com/2007/10/fact-checking-alan-greenspan.html.

p. 128 "Gentlemen, you are each entitled to your opinions

Variations of this quote have been attributed to American financier Bernard M. Baruch, Secretary of Defense James R. Schlesinger, and New York Senator Daniel Patrick Moynihan.

Chapter 7.
Silo Thinking: The Fourth Supermeme

p. 129 On September 9, 2009, two legends in science

Alvin Powell, "Wilson, Watson Reflect on Past Trials, Future Directions," *Harvard-Science*, September 10, 2009.

James Watson and Edward Wilson interviews at Sanders Theater, Harvard University, Cambridge, MA, September 9, 2009.

Interview with Dr. E. O. Wilson, Harvard University, Cambridge, MA, July 1, 2009.

p. 130 Watson insisted that more budget and resources

Steve Sailer, "James Watson as a Leader," Isteve.blogspot.com, October 22, 2007.

p. 130 Wilson's belief in the unification of knowledge

Edward O. Wilson, *Consilience* (New York: Knopf, 1998).

p. 132 Carol Kinsey Goman describes the effect of silo thinking

Carol Kinsey Goman, "Tearing Down Business 'Silos,'" www.sideroad.com.

Patrick Lencioni, *Silos, Politics, and Turf Wars: A Leadership Fable About Destroying the Barriers That Turn Colleagues into Competitors* (San Francisco: Jossey-Bass, 2006).

Hunter Hastings and Jeff Saperstein, *Bust the Silos: Opening Your Organization to Growth* (BookSurge.com, 2009).

"Collaborating Across Silos" (Boston: *Harvard Business Review*, July 6, 2009).

p. 132 According to the Institute of Medicine

www.healthypeople.gov/data/2010prog/focus01/.

p. 132 When the highest rates of readmission within the first thirty days

Stephen F. Jencks, M.D., M.P.H., Mark V. Williams, M.D., Eric A. Coleman, M.D., M.P.H., "Rehospitalizations among Patients in the Medicare Fee-for-Service Program," *New England Journal of Medicine*, April 2, 2009.

p. 133 Insurance companies were holding most of the cards

Joyce Frieden, "Coordinated Care Would Cut Medicare Readmissions," *Internal Medicine News*, October 15, 2009.

Danny Chun, "New System Reduces Hospital Readmissions for Congestive Heart Failure Patients," EurekAlert.org, November 27, 1996, http://www.scienceblog.com/community/older/1996/A/199600563.html.

Jennifer Silverman, "Malpractice Crisis Prompts More Referrals to ER: Instead of Office Treatment," *Pediatric News*, July 2005.

Mary Brophy, Marcus Bello, and Marisol Bello, "More Discharged Patients Are Returning via the ER," *USA Today*, September 9, 2009.

Michael Pistoria, "Readmissions: A Wake Up Call," *Family Practice News*, June 2009.

James J. Holloway and J William Thomas, "Factors Influencing Readmission Risk: Implications for Quality Monitoring," *Health Care Financing Review*, January 1, 1989.

Barbara Silliman, *You and the Broken American Healthcare System* (Bloomington, IN: Author House, 2008).

Selvoy M. Fillerup, *Chronic Crisis: Critical Care for America's Collapsing Healthcare System* (Gilbert, AZ: Acacia, 2007).

Lawrence Wolper, *Health Care Administration: Planning, Implementing, and Managing Organized Delivery Systems*, 4th ed. (Sudbury, MA: Jones and Bartlett, 2004).

Josei Marila Paganini, *Quality and Efficiency of Hospital Care: The Relationship Between Structure, Process, and Outcome* (Washington, DC: Pan American Health Organization, 1993).

p. 135 President Obama mentioned Anthem Blue Cross

Duke Helfand, "Congress Opens Investigation into Anthem Blue Cross," *Los Angeles Times*, February 10, 2010.

Stephanie Condon, "Obama Administration Blasts Anthem Blue Cross Rate Hikes," CBS News, February 8, 2010.

Erica Werner, "Anthem Asked to Justify Rate Hike in California," *San Francisco Examiner*, February 9, 2010.

p. 136 But in terms of quality it ranks thirty-seventh

Jim Peron, "Ranking the U.S. Healthcare System," Foundation for Economic Education, November 2007.

"The World Health Organization's Ranking of the World's Health Systems," www.geographic.org.

p. 137 In keeping with this, the CIA reinvigorated the MEDEA program

Saul Kaplan, "Innovators, Break Down Those Silos," *BusinessWeek*, February 8, 2010.

p. 137 It is shocking to discover how many independent nonprofit organizations are out to solve

"For the Common Good: The Economic Impact of Monterey County's Nonprofit Industry," Monterey County Board of Supervisors, Department of Social and Employment Services, meeting, April 21, 2009.

p. 138 Following the 2010 earthquake in Haiti, the amount of duplication between nonprofit silos

Anthony Boadle and Will Durham, "Doctors Group Complains Haiti Supplies Diverted," Reuters Alertnet, January 19, 2010.

Philip Dru, "Doctors Without Borders in Haiti: Why Couldn't They Land?" NWOTruth.com, January 18, 2010, http://nwotruth.com/doctors-without-borders-in-haiti-why-couldnt-they-land/.

Madhuri Dey, "Haiti Flight Logs Reveal Chaotic Supplies," Thaindian.com, February 19, 2010, http://www.thaindian.com/newsportal/world/haiti-flight-logs-reveal-chaotic-supplies_100323046.html.

p. 138 Dr. Carol Kinsey Goman, business consultant

Carol Kinsey Goman, "Tearing Down Business 'Silos,'" www.sideroad.com.

p. 139 Biologists who study chimpanzees

NOVA 01/01/07, "The Bonobo in All of Us," 9/29/2005. Interview conducted at the Columbus Zoo with Sue Western, scriptwriter for *Bonobo: Missing in Action* (the BBC version of *The Last Great Ape*), and edited by Peter Tyson, Editor in Chief of *NOVA Online*, http://www.pbs.org/wgbh/nova/beta/evolution/bonobo-all-us.html.

p. 139 Territoriality is the process by which

Adian Sammons, www.psychlotron.org.uk.

p. 140 From a historical perspective, as humans transitioned

Philip L. White and Michael L. White, "Why Do People Create Nationalities?" *Nation-*

ality, The History of a Social Phenomenon (Section B), pp. 83–104 (21), http://
nationalityinworldhistory.net/ch2B.html.

p. 140 "[I]n order to impose order and predictability on a complex"
Aidan Sammons, www.psychlotron.org.uk/resources/environmental/A2_OCR_env_
territory.pdf.

p. 141 NASA was signed into existence in 1958 by President Dwight D. Eisenhower
www.history.nasa.gov, www.jobmonkey.com/governmentjobs/work-for-nasa.html.

p. 142 So, NASA turned their attention to a new market
Interview with NASA chief scientist Dan Rasky, Carmel, CA, September 2009.

Brian Berger, "Report Urges U.S. to Pursue Space-Based Solar Power," Space.com, October 12, 2007, www.space.com/businesstechnology/071012-pentagon-space-solar-
power.html.

G. I. A. and I. Umarov, *Solar Energy* (NASA technical translation TTF-16, ISS, 1975).

Nick Allen, "NASA Launches Space-Based Solar Observatory," *Daily Telegraph*, February 11, 2010.

Committee of the Assessment of NASA's Space Solar Power Investment Strategy, Aeronautics Space Engineering Board, Division on Engineering and Physical Sciences
and the National Research Council, "Laying the Foundation for Space Solar Power:
An Assessment of NASA's Space Solar Power Investment Strategy" (Washington, DC:
National Academies Press, 2001).

Aeronautics and Space Engineering Board and the National Research Council, "Solar
Power Investment Strategy" (Washington, DC: National Academies Press, October
30, 2001).

NSF/NASA Solar Energy Panel, "An Assessment of Solar Energy as National Energy Resource," (January 1, 1972).

Terrestrial Energy Generation Based on Space Solar Power: A Feasible Concept or Fantasy?
(workshop sponsored by the MIT Technology and Development Program, Cambridge, MA, May 14–16, 2007).

Jonathan Marshall, "Space Solar Power: The Next Frontier?" April 13, 2009, www.
next100.com/2009/04/space-solar-power-the-next-fro.php.

"Japan to Beam Solar Power from Space on Lasers," *Fox News*, November 9, 2009.

Peter E. Glaser, "Power from the Sun: Its Future," *Science*, November 22, 1968.

P. E. Glaser, O. E. Maynard, J. Mackovciak, and E. L. Ralph, *Feasibility Study of a Satellite Solar Power Station* (Cambridge, MA: Arthur D. Little, February 1974).

John C. Mankins, "A Fresh Look at Space Solar Power: New Architectures, Concepts,
and Technologies," IAF paper no. IAF-97-R.2.03, http://www.spacefuture.com/
archive/a_fresh_look_at_space_solar_power_new_architectures_concepts_and_
technologies.shtml.

W. C. Brown, "The History of Power Transmission by Radio Waves," *IEEE Transactions
on Microwave Theory and Techniques* 32(9) (September 1984).

N. M. Komerath and N. Boechler, "The Space Power Grid" (paper presented at the Fifty-
seventh International Astronautical Federation Congress, Valencia, Spain, October
2006).

N. Shinohara, "Wireless Power Transmission for Solar Power Satellite" (Space Solar
Power Workshop, Georgia Institute of Technology, Atlanta, GA).

Rice University, comp., *Solar Power Satellite Offshore Rectenna Study*, Final Report
(Houston. Rice University, November 1980)

"Researchers Beam 'Space' Solar Power in Hawaii," September 12, 2008, www.
wired.com/wiredscience/2008/09/visionary-beams/.

p. 142 Electricity from satellites in outer space?
Michael D. Lemonick, "Solar Power from Space: Moving Beyond Science Fiction," *Yale Environment 360*, August 31, 2009.

p. 144 In his 1998 book *Consilience: The Unity of Knowledge*
Edward O. Wilson, *Consilience* (New York: Knopf, 1998).

p. 145 Here writer Saul Kaplan gets the final word
Saul Kaplan, "Innovators, Break Down Those Silos," *BusinessWeek*, February 8, 2010.

Chapter 8.
Extreme Economics: The Fifth Supermeme

p. 149 Kamen sees opportunities for improvement everywhere
Brian Braiker, "Big Problem, Neat Solution," *Newsweek*, April 5, 2008.
Isabella Rossellini interview with Dean Kamen, *Iconoclasts*, Sundance Channel, November 16, 2006.
Allan J. Organ, *The Regenerator and the Stirling Engine* (New York: Wiley, 1997).
William R. Martini, *Stirling Engine Design Manual* (Honolulu: University Press of the Pacific, 2004).
Theodore Finkelstein and Allan Organ, *Air Engines: The History, Science, and Reality of the Perfect Engine* (New York: ASME Press, 2004).
Matthew R. Freije, *Disinfecting Potable Water Systems* (Solana Beach, CA: HC Information Resources, 2004).
Terrold J. Troyan and Sigurd P. Haber, *Treatment of Microbial Contaminants in Potable Water Supplies: Technologies and Costs* (Norwich, NY: William Andrew, 1991).

p. 151 Just ask solar panel manufacturers
Reece Ray, *The Sun Betrayed* (Boston: South End Press, 1999).
Michael Silverstein, *The Once and Future Resource: A History of Solar Energy* (Environmental Design and Research Center, 1977).
Karl W. Boer, *The Fifty Year History of the International Solar Energy Society* (Boulder, CO: American Solar Energy Society, 2005).
Solar Energy Institute, *Webster's Timeline History, 1979–2000* (San Diego: Icon Group International, 2009).
Raul Lopez-Aguilar, Bernardo Murillo-Amador, and Gualalupe Rodriguez-Quezada, *Hydroponic Green Fodder (HFG): An Alternative for Cattle Food Production in Arid Zones* (Recife, Brazil: Interciencia Association, September 16, 2009).
"IFF Breaks New Ground in Hydroponics Research," *Household and Personal Products Industry*, July 28, 2005.

p. 152 We may not always use the same language as an economist
R. W. Belk and M. Wallendorf, "The Sacred Meanings of Money," *Journal of Economic Psychology* 11 (1999).
Herb Goldberg and Robert T. Lewis, *Money Madness: The Psychology of Saving, Spending, Loving, and Hating Money* (Issaqua, WA: Wellness Institute, January 2000).
Michael Argyle and Adrian Furnham, *The Psychology of Money* (New York: Routledge, 1998).
Cele C. Otnes and Tina M. Lowrey, eds., *Contemporary Consumption Rituals: A Research Anthology* (Mahwah, NJ: Lawrence Erlbaum, 2004).
C. B. Burgogyne and D. A. Routh, *Journal of Economic Psychology*, 1991.
Nigel Dodd, *The Sociology of Money: Economics, Reason, and Contemporary Society* (New York: Continuum, 1994).

Paul W. Glimcher, *Decisions, Uncertainty, and the Brain: The Science of Neuroeconomics* (Cambridge, MA: MIT Press, 2003).

Richard H. Thaler, *The Winner's Curse: Paradoxes and Anomalies of Economic Life* (Princeton, NJ: Princeton University Press, 1992).

Viviana A. Rotman Zelizer, *The Social Meaning of Money: Pin Money, Pay Checks, Poor Relief, and Other Currencies* (Princeton, NJ: Princeton University Press, 1997).

B. G. Carruthers and W. N. Espeland, "Money, Meaning, and Morality," *American Behavioral Scientist* 41(10) (1998).

Thomas Crump, *The Phenomenon of Money* (London: Routledge and Kegan Paul, 1981).

Marc Shell, *Money, Language, and Thought* (Berkeley and Los Angeles: University of California Press, 1982).

Philip E. Slater, *Wealth Addiction* (New York: Dutton, 1980).

p. 152 Prenups are modern contracts designed to define the financial terms of marriage

Lisa Smith, "Marriage, Divorce, and the Dotted Line," www.investopedia.com/articles/pf/06/prenuptialagreement.asp.

Jan Pahl, *Money and Marriage* (New York: Macmillan, 1989).

p. 153 H&M stores were throwing large bags of brand new, unsold goods

Jim Dwyer, "A Clothing Clearance Where More Than Just the Prices Have Been Slashed," *New York Times*, January 5, 2010.

Interview with Dr. E. O. Wilson, Harvard University, Cambridge, MA, July 1, 2009.

p. 155 A few years ago, graduate students at a major university devised an experiment

Presentation by Steven D. Levitt and Stephen J. Dubner at Yahoo headquarters, Sunnyvale, CA, July 28, 2005.

Sarah F. Brosnan, Mark F. Grady, Susan P. Lambeth, Steven J. Schapiro, and Michael J. Beran, "Chimpanzee Autarky," *PLoS ONE* 3(1) (December 29, 2007).

E. G. Lea and Paul Webley, *Money as Tool, Money as Drug: The Biological Psychology of a Strong Incentive* (Cambridge: Cambridge University Press, 2005).

K. G. Duffy, R. W. Wrangham, and J. B. Silk, "Male Chimpanzees Exchange Political Support for Mating Opportunities," *Current Biology* 17(15) (2007).

Jeffrey R. Stevens, "The Selfish Nature of Generosity: Harassment and Food Sharing in Primates," *The Royal Society*, October 29, 2003.

S. F. Brosnan and F. B. M de Waal, *Journal of Comparative Psychology* 118 (2004).

C. W. Hyatt and W. D. Hopkins, *Behavioral Processes* (N.p.: Elsevier, 1998).

Frans de Waal, *Chimpanzee Politics: Power and Sex Among Apes* (London: Jonathan Cape, 1982).

V. Dufour, E. H. M. Sterk, M. Pele, and B. Thierry, "Chimpanzee (Pan Troglodytes) Anticipation of Food Return: Coping with Waiting Time in an Exchange Task," *Journal of Comparative Psychology* 121(2) (2007).

S. F. Brosnan and F. B. M. de Waal, "A Simple Ability to Barter in Chimpanzees, Pan Troglodytes," *Primates* 46(3) (2005).

S. F. Brosnan and F. B. M. de Waal, "Monkeys Reject Unequal Pay," *Nature* 425 (2003).

C. Sousa and T. Matsuwasa, "The Use of Tokens as Rewards and Tools by Chimpanzees (Pan Troglodytes)," *Animal Cognition* 4(3–4) (2001).

S. F. Brosnan and F. B. M. de Waal, "Socially Learned Preferences for Differentially Rewarded Tokens in the Brown Capuchin Monkey, Cebus Apella," *Journal of Comparative Psychology* 118(2) (2004).

p. 158 Richard Dawkins provides an evolutionary explanation

Richard Dawkins, *The Selfish Gene* (New York: Oxford University Press, 2006).

John R. Krebs and Richard Dawkins, *Animal Signals: Mind Reading and Manipulation*, in *Behavioral Ecology: An Evolutionary Approach*, ed. John R. Krebs and Nicholas B. Davies, 2nd ed. (Oxford: Blackwell, 1984).

Kenneth E. Boulding, *Evolutionary Economics* (Thousand Oaks, CA: Sage, 1981).

Robert H. Frank, "If 'Homoeconomicus' Could Choose His Own Utility Function, Would He Want One with a Conscience?" *American Economic Review* 79 (June 1989).

Kurt Dopfer, ed., *The Evolutionary Foundation of Economics* (Cambridge: Cambridge University Press, 2006).

Michael Shermer, *The Mind of the Market: How Biology and Psychology Shape Our Economic Lives* (New York: Holt Paperbacks, 2006).

J. Stanley Metcalfe, *Evolutionary Economics and Creative Destruction* (New York: Routledge, 1998).

Arthur Gandolfi, Anna Gandolfi, and David Barash, *Economics as an Evolutionary Science: From Utility to Fitness* (Edison, NJ: Transaction, 2002).

Daniel Friedman, *Morals and Markets: An Evolutionary Account of the Modern World* (New York: Palgave Macmillan, 2008).

Jason Potts, *The New Evolutionary Microeconomics: Complexity, Competence, and Adaptive Behavior* (Cheltenham, UK: Edward Elgar, 2001).

Peter Koslowski, *Sociobiology and Bioeconomics: The Theory of Evolution in Biological and Economic Theory* (New York: Springer, 1999).

Sunny Y. Auyand, *Foundations of Complex-System Theories in Economics, Evolutionary Biology, and Statistical Physics* (Cambridge: Cambridge University Press, 1999).

Charles Darwin, *The Descent of Man* (New York: Penguin, 2007).

Charles Darwin, *The Expression of the Emotions in Man and Animals* (New York: Oxford University Press, 1998).

Chris Stringer and Peter Andrews, *The Complete World of Human Evolution* (New York: Thames and Hudson, 2005).

Gregory Cochran and Henry Harpending, *The 10,000 Year Explosion: How Civilization Accelerated Human Evolution* (New York: Basic Books, 2009).

John Cartwright, *Evolution and Human Behavior: Darwinian Perspectives on Human Nature* (Mendham, UK: Bradford/MIT Press, 2000).

Eva Jablonka and Marion J. Lamb, *Evolution in Four Dimensions: Genetic, Epigenetic, Behavioral, and Symbolic Variation in the History of Life* (Cambridge, MA: MIT Press, 2005).

Michael J. Behe, *The Edge of Evolution: The Search for the Limits of Darwinism* (New York: Free Press, 2008).

Christopher Scarre, *The Human Past: World Prehistory and the Development of Human Societies* (New York: Thames and Hudson, 2005).

Timothy Goldsmith and William F. Zimmerman, *Biology, Evolution, and Human Nature* (New York: Wiley, 2000).

Philip Clayton and Jeffrey Schloss, *Evolution and Ethics: Human Morality in Biological and Religious Perspective* (Grand Rapids, MI: Eerdmans, 2004).

Bruce H. Lipton and Steve Bhaerman, *Spontaneous Evolution: Our Positive Future (and a Way to Get There from Here)* (Carlsbad, CA: Hay House, 2009).

Richard Restak, *The New Brain: How the Modern Age Is Rewiring Your Mind* (Emmaus, PA: Rodale, 2004).

Mihail C. Roco and Carlo D. Montemagno, *The Coevolution of Human Potential and Converging Technologies* (New York: New York Academy of Science, 2004).

Robert J. Sternberg and Janet E. Davidson, *The Nature of Insight* (Cambridge, MA: MIT Press, 1995).

p. 159 Harvard researcher Terence Charles Burnham states it another way

Terry Burnham and Jay Phelan, *Mean Genes: From Sex to Money to Food: Taming Our Primal Instincts* (New York: Perseus, 2000).

Terence Burnham and Brian Hare, *Engineering Human Cooperation: Does Involuntary Neural Activation Increase Public Goods Contribution?* (self-published, June 2005).

Terence C. Burnham, "Essays on Genetic Evolution and Economics" (Ph.D. diss., Committee of Business Economics, Harvard University, 1997), www.bookpump.com/dps/pdf-b/5856429b.pdf.

S. E. G. Lea, Roger M. Tarpy, and Paul Webley, *The Individual in the Economy: A Textbook of Economic Psychology* (Cambridge: Cambridge University Press, 1987).

John Foster and Werner Holzi, *Applied Evolutionary Economics and Complex Systems* (Cheltenham, UK: Edward Elgar, 2004).

p. 159 But to truly understand the origins of the economics supermeme

Niall Ferguson, *The Ascent of Money: A Financial History of the World* (New York: Penguin, 2008).

Carol Schwalberg, *From Cattle to Credit Cards: The History of Money* (New York: Meredith Press, 1969).

John Kenneth Galbraith, *Money: Whence It Came, Where It Went* (Boston: Houghton Mifflin, 2001).

Glyn Davies, *A History of Money: From Ancient Times to the Present Day* (Cardiff, UK: University of Wales Press, 2002).

William N. Goetzmann, *Financing Civilization*, http://viking.som.yale.edu/will/finciv/chapter1.htm.

"The History of Credit," Myvesta.org/history.

Carl Menger, "The Origin of Money" (Greenwich, CT: Committee for Monetary Research and Education, 1984).

Charles F. Horne, *The Code of Hammurabi* (Forgotten Books, 2007).

p. 162 For every $135 the average American saved

Leah Theis, *United States Census Bureau, Statistical Abstracts, 651: Relation of GDP, GNP, Net National Product, National Income, Personal Income, Disposable Personal Income, and Personal Savings* (Washington, DC: U.S. Census Bureau, 2008).

Federal Reserve, "Consumer Credit" (Washington, DC: Federal Reserve Statistical Release, September 8, 2008).

Mark Brinker, "Credit Card Debt Statistics" (Clinton Township, MI: Hoffman, Brinker and Roberts, August 2008), www.hoffmanbrinker.com/credit-card-debt-statistics.html.

Ben Woolsey and Matt Schulz, "Credit Card Industry Facts, Debt Statistics 2006–2008," Creditcards.com, 2008, www.creditcards.com/credit-card-news/credit-card-industry-facts-personal-debt-statistics-1276.php.

Lloyd Klein, *It's in the Cards: Consumer Credit and the American Experience* (Westport, CT: Praeger, 1999).

Robert D. Manning, *Credit Card Nation: The Consequences of America's Addiction to Credit* (New York: Basic Books, 2001).

Matty Simmons, *The Credit Card Catastrophe: The 20th Century Phenomenon That Changed the World* (Ft. Lee, NJ: Barricade Books, 1995).

Robert H. Scott III, "Credit Card Use and Abuse: A Veblenian Analysis." *Journal of Economic Issues*, June 13, 2007.

Lawrence M. Ausubel, "Credit Card Default, Credit Card Profits, and Bankruptcy," *American Bankruptcy Law Journal*, Spring 1997.

"U.S. Savings Rate Hits Lowest Level Since 1933," Associated Press, January 30, 2006.

Donna Boundy, *When Money Is the Drug: The Compulsion for Credit, Cash, and Chronic Debt* (New York: Harper, 1993).

T. Newton, "Credit and Civilization," *British Journal of Sociology* 54(3) (2003).

p. 162 In 2010, the federal debt in the United States

United States National Debt Clock, http//www.brillig.com/debt-clock/.

Shayne C. Kavanaugh, "Examining the Increasing Level of Federal Debt," *Government Finance Review*, April 3, 2009.

Brian W. Cashell, "The Federal Government Debt: Its Size and Economic Significance" (Washington, DC: Congressional Research Service, July 7, 2009).

William Bonner and Addison Wiggin, *The New Empire of Debt: The Rise and Fall of an Epic Financial Bubble*, 2nd ed. (Hoboken, NJ: Wilcy, 2009).

Robert E. Wright, *One Nation Under Debt: Hamilton, Jefferson, and the History of What We Owe* (New York: McGraw-Hill, 2008).

Andrew L. Yarrow, *Forgive Us Our Debt: The Intergenerational Dangers of Fiscal Irresponsibility* (New Haven, CT: Yale University Press, 2008).

Donna Boundy, *When Money Is the Drug: The Compulsion for Credit, Cash, and Chronic Debt* (New York: Harper, 1993).

p. 163 Now add in the deficits accumulated by individual states

Kevin Hassett, "D Is for Deficit: Guess Who's to Blame for State Budget Problems?" *Wall Street Journal*, October 6, 2003.

"State Surplus or Deficit per Household, Fiscal Year 2006" (Washington, DC: The Tax Foundation).

Elizabeth Hill, "2008–09 Overview of the Governor's Budget" (Sacramento: Legislative Analyst's Office, January 14, 2008).

Frank Keegan, "Economists: State Deficits Could Stall Recovery," Watchdog.com, November 12, 2009, http://watchdog.org/1486/economists-state-deficits-could-stall-recovery/.

Colin Berr, "More and More States on Budget Brink," CNNMoney, January 15, 2010.

Sarah Burrows, "Of 36 States Facing Deficits This Year, 22 Are Increasing Spending," CNS News, December 17, 2008, http://www.cnsnews.com/news/article/40908.

Donna Boundy, *When Money Is the Drug: The Compulsion for Credit, Cash, and Chronic Debt* (New York: Harper, 1993).

p. 164 *Then someone suggested legalizing marijuana*

http://personalmoneystore.com/moneyblog/2009/02/25/legalizing-marijuana-fix-californias-economy.

Alison Stateman, "Can Marijuana Help Rescue California's Economy?" *Time*, March 13, 2009.

Elizabeth Fairchild, "Could Legalizing Marijuana Fix California's Economy?" *Personal Money Store*, February, 25, 2009.

Joyce D. Henry, "Senator T. Milton Strett's Tax Plan for Uniform Commercial Crops of Marijuana" (1983).

Kenneth W. Clements and Xueyan Zhao, *Economics and Marijuana: Consumption, Pricing, and Legalisation* (Cambridge: Cambridge University Press, 2009).

Lisa Leff, "Legalizing Marijuana Would Cause Prices to Plummet," *Huffington Post*, July 7, 2010.

p. 167 There is more attention paid to raising capital

Jennifer Washburn, *University, Inc.: The Corporate Corruption of Higher Education* (New York: Basic Books, 2006).

Derek Curtis Bok, *Universities in the Marketplace: The Commercialization of Higher Education* (Princeton, NJ: Princeton University Press, 2004).

Burton H. Weisbrod, Jeffrey P. Bailou, and Evelyn D. Asch, *Mission and Money: Understanding the University* (New York: Cambridge University Press, 2008).

Condoleeza Rice, "Rice Outlines Budget Pressures for Fiscal Year 1998," Stanford News Service, December 10, 1996.

David Stauth, "Budget Pressures Raise Research Funding Concerns," *Oregon State University News and Communications,* February 1997.

Robert Reinhold, "Budget Cuts Jar University of California," *New York Times,* January 21, 1991.

Justin Harris, "Donations to Universities May Decline in Next Two Years," *The State News,* March 16, 2009.

Aldo Geuna, *The Economics of Knowledge Production: Funding and Structure of University Research* (Cheltenham, UK: Edward Elgar, 1999).

James D. Savage, *Funding Science in America: Congress, Universities, and the Politics of Academic Pork Barrel* (Cambridge: Cambridge University Press, 2000).

Becky Gillette, "As Public Funding Erodes, Universities Change," *Mississippi Business Journal,* July 14, 2006.

M. Gulbrandsen and J. C. Smeby, "Industry Funding and University Professor's Research Performance" *Research Policy* 34(6) (August 2005).

"Non-Federal R&D Funding for U.S. Universities Increases," *Instrument Business Outlook,* November 16, 2009.

David L. Kirp, Elizabeth Popp Berman, Jeffrey T. Holman, and Patrick Roberts, *Shakespeare, Einstein, and the Bottomline: The Marketing of Higher Education* (Cambridge, MA: Harvard University Press, 2004).

John C. Knapp and David J. Siegel, eds., *The Business of Higher Education* (Santa Barbara: Praeger, 2009).

p. 167 There was "an intensified need for collaboration"

American Society for Engineering Education, comp., "Intellectual Property: Universities, Corporations, and Finding a Common Ground," February 13, 2006, www.asee.org/activities/organizations/councils/edc/2006-IP-White-Paper/IPWhite Paper-WEB.pdf.

p. 170 Pharmaceutical profits were five and a half times greater

Neal Pattison and Luke Warren, principal authors, under the direction of Ben Peck and Frank Clemente, "Drug Industry Profits: Hefty Pharmaceutical Company Margins Dwarf All Other Industries" (Washington, DC: Public Citizen's Congress Watch, 2003), www.citizen.org/documents/Pharma_Report.pdf.

Linda Marsa, *Prescription for Profits: How The Pharmaceutical Industry Bankrolled the Unholy Marriage Between Science and Business* (New York: Scribner, 1997).

Scott Hensley, "Follow the Money: Drug Prices Rise at a Faster Clip, Placing Burden on Consumers," *Wall Street Journal,* April 15, 2003.

Leonard Weber, *Profits Before People: Ethical Standards and the Marketing of Prescription Drugs* (Bloomington: Indiana University Press, 2006).

Duncan Reekie and Michael H. Weber, *Profits, Politics, and Drugs* (Teaneck, NJ: Holmes and Meier, 1979).

p. 171   One day Microsoft and Google are on the verge of joining forces,

Jon Swartz and Michelle Kessler, "Microsoft, Google Cook Up Deal," San Francisco, November 3, 2003.

John Markoff and Andrew Ross Sorkin, "Microsoft and Google—Partners or Rivals?" *New York Times*, October 31, 2003.

p. 173   I have a difficult time keeping track of all the different religious sects

"List of Known Terrorist Organizations" (Washington, DC: Center for Defense Information, 2009).

"National Strategy for Combating Terrorism" (Washington, DC: U.S. Department of State, February 2003).

"Country Reports on Terrorism" (Washington, DC: U.S. Department of State, Office of the Coordinator for Counterterrorism, April 2003).

"28 Groups on the U.S. Department of State's Designated Foreign Terrorist Organizations List," www.fbi.gov/publications/terror/terrorism2002_2005.htm.

Audrey Cronin, "The FTO List and Congress: Sanctioning Designated Foreign Terrorist Organizations" (Washington, DC: Congressional Research Service, December 30, 2009).

p. 174   In his book *The Stillborn God*

Mark Lilla, *The Stillborn God: Religion, Politics, and the Modern West* (New York: Vintage, 2008). See also Lilla's article at www.nytimes.com/2007/08/19/magazine/19Religion-t.html?pagewanted=1&_r=1.

p. 176   Imagine the confusion in the White House

Mark Lilla, "The Politics of God," *New York Times Magazine*, August 19, 2007.

p. 177   Is an expert at "Islamic Economics"

Timur Kuran, "The Genesis of Islamic Economics: A Chapter in the Politics of Muslim Identity," *Social Research* 64(2) (Summer 1997).

Sohrab Behdad, "Property Rights in Contemporary Islamic Economic Thought: A Critical Perspective," *Review of Social Economy* 47(2) (1989).

Sean S. Costigan and David Gold, eds., *Terrornomics* (Burlington, VT: Ashgate, 2007).

Eli Berman, *Radical, Religious, and Violent: The New Economics of Terrorism* (Cambridge, MA: MIT Press, 2009).

Muhammad Abdul-Rauf, *A Muslim's Reflections on Democratic Capitalism* (Washington, DC: American Enterprise Institute, February 1984).

M. Umer Chapra, *Islam and the Economic Challenge* (Leicester, UK: Islamic Foundation, 1992).

Timur Kuran, "The Economic System in Contemporary Islam Thought: Interpretation and Assessment," *International Journal of Middle Eastern Studies* 18 (1986).

Sayyid Abu'l-A'la Mawdudi, *The Economic Problem of Man and Its Islamic Solution* (Lahore, Pakistan: Islamic Publications, 1978).

Sayyid Abu'l-A'la Mawdudi, "The Rudiments of Islamic Philosophy of Economics," in *Selected Speeches and Writings of Mawlana Mawdudi*, vol. 1, trans. S. Zakir Aijaz (Karachi, Pakistan: International Islamic Publishers, 1981).

Vikas Mishra, *Hinduism and Economic Growth* (Cambridge: Cambridge University Press, 1962).

Maxine Rodinson, *Islam and Capitalism* (New York: Pantheon, 1973).

Daryush Shayegan, *Cultural Schizophrenia: Islamic Societies Confronting the West* (London: Sagi, 1992).

Muhammad Nejatullah Siddiqui, *Muslim Economic Thinking: A Survey of Contemporary Literature* (Markfield, UK: Islamic Foundation, 1981).

Timur Kuran, "The Logic of Financial Westernization in the Middle East," *Journal of Economic Behavior and Organization* 59 (April 2005).

Mohamed Aslam Haneef, *Contemporary Islamic Economic Thought: A Selected Comparative Analysis* (Kuala Lumpur, Malaysia: Ikraq, 1995).

Timur Kuran, "On the Notion of Economic Justice in Contemporary Islamic Thought," *International Journal of Middle East Studies* 21 (1989).

Syed Nawab Haider Naqvi, *Islam Economics and Society* (London: Kegan Paul International, 1994).

Alan Richards and John Waterbury, *A Political Economy of the Middle East* (Boulder, CO: Westview Press, 1996).

Muhammad Baqir Al-Sadr, *Iqtsaduna: Our Economics* (Tehran, Iran: Tehran World Organization for Islamic Services, 1982).

Timur Kuran, "Islamic Economics and Islamic Subeconomy," *Journal of Economic Perspectives* (Fall 1995).

Timur Kuran, "Religious Economics and the Economics of Religion," *Journal of Institutional and Theoretical Economics*, December 1994.

Conversation with Timur Kuran, *USC Magazine*, December 18, 2006.

Timur Kuran, "Islam and Economic Underdevelopment: An Old Puzzle Revisited," *Journal of Institutional and Theoretical Economics*, March 1997.

p. 177 A new age bible for business tycoons

Ayn Rand, *Atlas Shrugged* (New York: Penguin, 1999).

p. 178 Albert Camus characterized the role rational thought plays

www.wisdomquotes.com.

Chapter 9.
Surmounting the Supermemes:
Rational Solutions in an Irrational World

p. 179 Hawken had spent a great deal of his life fighting modern supermemes

Paul Hawken, "The Class of 2009 Commencement Address," University of Portland, Portland, Oregon, May 3, 2009, official transcript, http://www.up.edu/commencement/default.aspx?cid=9456.

Paul Hawken, *Blessed Unrest: How the Largest Movement in History Is Restoring Grace, Justice, and Beauty to the World* (New York: Penguin, 2008).

p. 180 In 1633 Galileo was sentenced by the Catholic church

Mike Price, "Galileo, Reconsidered," *Smithsonian*, August 12, 2008.

Bertolt Brecht, *Life of Galileo* (New York: Penguin, 2008).

David Brewster, *The Martyrs of Science, or the Lives of Galileo, Tycho Brahe, and Kepler* (Whitefish, MT: Kessinger, 1844).

Andrea Frova and Mariapiera Marenzana, *Thus Spoke Galileo: The Great Scientist's Ideas and Their Relevance to the Present Day*, trans. James H. McManus (New York: Oxford University Press, 2006).

Hal Hellman, *Great Feuds in Science: Ten of the Liveliest Disputes Ever* (New York: Wiley, 1999).

Thomas S. Kuhn, *The Structure of Scientific Revolutions* (Chicago: University of Chicago Press, 1996).

Jim Sidanius and Felicia Pratto, "The Inevitability of Oppression and the Dynamics of Social Dominance," in *Prejudice, Politics, and the American Dilemma*, ed. Paul Sniderman, Philip E. Tetlock, and Edward C. Carmines (Palo Alto, CA: Stanford University Press, 1993).

p. 180 Charles Darwin was also reluctant to publish

Dorothy Hinshaw Patent, *Charles Darwin: The Life of a Revolutionary Thinker* (New York: Holiday House, 2001).

David Quammen, *The Reluctant Mr. Darwin: An Intimate Portrait of Charles Darwin and the Making of His Theory of Evolution* (New York: Norton, 2006).

Charles Van Doren, *A History of Knowledge: Past, Present, and Future* (New York: Ballantine Books, 1982).

Lyanda Lynn Haupt, *Pilgrim on the Great Bird Continent: The Importance of Everything and Other Lessons from Darwin's Lost Notebooks* (New York: Little, Brown, 2006).

Michael Ruse, *Charles Darwin* (Hoboken, NJ: Wiley-Blackwell, 2008).

Laura Fermi and Gilberto Bernardini, *Galileo and the Scientific Revolution* (New York: Basic Books, 1961).

Robert J. Sternberg and Janet E. Davidson, *The Nature of Insight* (Cambridge, MA: MIT Press, 1995).

Jim Sidanius and Felicia Pratto, "'The Inevitability of Oppression and the Dynamics of Social Dominance," in *Prejudice, Politics, and the American Dilemma*, ed. Paul Sniderman, Philip E. Tetlock, and Edward C. Carmines (Palo Alto, CA: Stanford University Press, 1993).

p. 180 Public humiliation that Martin Fleischmann and Stanley Pons endured

Steven Krivit, "The Cold Fusion Short Story," *New Energy Times*, January 5, 2007.

Malcolm W. Browne, "Physicists Debunk Claim of a New Kind of Fusion," *New York Times*, May 3, 1989.

"Cold Fusion Is Hot Again," *60 Minutes*, April 19, 2009, http://www.cbsnews.com/stories/2009/04/17/60minutes/main4952167.shtml?tag=contentMain;contentBody.

Jim Sidanius and Felicia Pratto, "The Inevitability of Oppression and the Dynamics of Social Dominance," in *Prejudice, Politics, and the American Dilemma*, ed. Paul Sniderman, Philip E. Tetlock, and Edward C. Carmines (Palo Alto, CA: Stanford University Press, 1993).

p. 181 Warren Buffett, worth $42 billion, publicly announced he would begin giving away

Carol J. Loomis, "Warren Buffett Gives Away His Fortune," *Fortune*, June 25, 2006.

Alessandro Della Bella, "Bill Gates' Foundation Pledges $10 Billion for Vaccines," Associated Press, January 29, 2010.

Michael Kinsey and Conor Clarke, *Creative Capitalism: A Conversation with Bill Gates, Warren Buffett, and Other Economic Leaders* (New York: Simon and Schuster, 2009).

Joel L. Fleishman, *The Foundation: A Great American Secret; How Private Wealth Is Changing the World* (New York: PublicAffairs, 2009).

Marc Benioff and Carlye Adler, *The Business of Changing the World: Twenty Great Leaders on Strategic Corporate Philanthropy* (New York: McGraw-Hill, 2006).

p. 181 In 2006, Muhammad Yunus won the coveted Nobel Peace Prize

Muhammad Yunus, *Creating a World Without Poverty: Social Business and the Future of Capitalism* (New York: PublicAffairs, 2007).

"Profile: Dr. Muhammad Yunus," *Bangladesh News*, October 14, 2006.

Muhammad Yunus, *Banker to the Poor: The Autobiography of Muhammad Yunus* (Lake Havasu City, AZ: London Bridge, 2000).

"Introduction to Grameen Bank," Grameen-infor.org, February 11, 2010, www.grameen-info.org/index.php?option=com_content&task=view&id=26&Itemid=175.

"'Banker to the Poor' Gives New York Woman a Boost," Reuters, April 23, 2006.

Stefan Lovegren, "Nobel Peace Prize Goes to Micro-Loan Pioneers," *National Geographic News*, October 13, 2006.

p. 183 What's more, Yunus reports that "since it opened, the bank has given out"

Muhammad Yunus, *Creating a World Without Poverty: Social Business and the Future of Capitalism* (New York: PublicAffairs, 2007).

p. 183 According to the article "Small Loans Empower"

Sue Wheat, "Small Loans Empower," grameen-info.org, www.grameen-info.org/dialogue/dialogue31/small loan.htm.

Sue Wheat, "The Future of Microfinance: Banking the Unbankable" (report, London: Panos, 1997).

Beatriz Armendariz and Jonathan Morduch, *The Economics of Microfinance* (Cambridge, MA: MIT Press, 2007).

Jurriaan Kamp, *Small Change: How Fifty Dollars Can Change the World* (New York: Cosimo, 2006).

C. K. Prahalad, *The Fortune at the Bottom of the Pyramid: Eradicating Poverty Through Profits* (Upper Saddle River, NJ: Wharton School, 2006).

Alex Counts, *Small Loans, Big Dreams: How Nobel Prize Winner Muhammad Yunus and Microfinance Are Changing the World* (Hoboken, NJ: Wiley, 2008).

Suresh Sundaresan, ed., *Microfinance: Emerging Trends and Challenges* (Cheltenham, UK: Edward Elgar, 2009).

p. 184 He was unable to find financial support

Muhammad Yunus, *Creating a World Without Poverty: Social Business and the Future of Capitalism* (New York: PublicAffairs, 2007), pp. 47–48.

p. 186 Yunus's insight was so successful

Muhammad Yunus, *Creating a World Without Poverty: Social Business and the Future of Capitalism* (New York: PublicAffairs, 2007), p. 51.

Grameen Bank, official Web site: http://www.grameen.com/index.php?option=com_content&task=view&id=26&Itemid=175.

p. 186 Contrast this against the most successful banks

Douglas V. Gnazzo, "Money Part VIII: Fractional Reserve Lending," May 10, 2006, www.321gold.com/editorials/gnazzo/gnazzo051906.html.

Joshua N. Feinman, Jana Deschler, and Christoph Hinkelmann, "Reserve Requirements: History, Current Practice, and Potential Reform," *Federal Reserve Bulletin*, June 1, 1993.

Frederic P. Miller, Agnes F. Vandome, and John McBrewster, *Fractional Reserve Banking* (Beau Bassin, Mauritius: Alphascript, 2009).

Z. Nuri, "Fractional Reserve Banking as Economic Parasitism: A Scientific, Mathematical, and Historical Expose, Critique, and Manifesto," (2008).

Board of Governors of the Federal Reserve System, "The History of Reserve Requirements in the United States," *Federal Reserve Bulletin* 25 (November 1938).

NY Fed, official Web site: http://www.newyorkfed.org/aboutthefed/fedpoint/fed45.html

p. 186 By rejecting the idea of *profit at any cost*

Federal Reserve, "Consumer Credit" (Washington, DC: Federal Reserve Statistical Release, September 8, 2008).

Mark Brinker, "Credit Card Debt Statistics" (Clinton Township, MI: Hoffman, Brinker and Roberts, August 2008).

Ben Woolsey and Matt Schulz, "Credit Card Industry Facts, Debt Statistics 2006–2008," Creditcards.com, 2008, www.creditcards.com/credit-card-news/credit-card-industry-facts-personal-debt-statistics-1276.php.

Lucia F. Dunn and Taehyung Kim, "An Empirical Investigation of Credit Card Default" (Columbus: Department of Economics, Ohio State University, August 1999).

Chapter 10.
Awareness and Action: A Tactical Approach

p. 187 Many years ago when I was traveling in Japan
"I Am Awake," www.sinc.sunysb.edu.

p. 190 And because product lifecycles in Silicon Valley are short-lived
Pallab Chatterjee, "Short Product Life Cycles Demand Innovation Through Business Supply Chain Leader."

Fairborz Ghadar, "Shorter Product Life Cycles Dictate New Global Marketing Rules" (University Park: Pennsylvania State Smeal Center for Global Business Studies).

Robert G. Cooper, *Winning at New Products: Accelerating the Process from Idea to Launch* (New York: Basic Books, 2001).

Dave Brock, "Technology Companies Must Follow the Fashion Leaders!" (Mission Viejo, CA: Partners in Excellence, 2008).

Clayton M. Christensen, *The Innovator's Dilemma: The Revolutionary Book That Will Change the Way You Do Business* (New York: Harper Paperbacks, 2010).

Art Bell, *The Quickening: Today's Trends, Tomorrow's World* (New Orleans: Paper Chase, 1998).

Martin Ford, *The Light's in the Tunnel: Accelerating Technology and the Economy of the Future* (CreateSpace, 2009).

Michael J. Mauboussim and Alexander Schay, "Innovations and Markets: How Innovation Affects the Investing Process," for Credit Suisse First Boston Corporation, December 12, 2000.

F. G. Patterson, Jr., "Life Cycles for System Acquisition," George Mason University, Fairfax, VA.

p. 191 There are many reasons that short-term mitigations fail
Barry Leonard, "Developing the Mitigation Plan: Identifying Mitigation Actions and Implementing Strategies," October 31, 2003, http://www.eeri.org/mitigation/resource-library/policy-and-community-planners/fema-localstate-guides/developing-the-mitigation-plan-identifying-mitigation-actions-and-implementing-strategies-fema-386-3.

Stephen S. Benham, *Actionable Strategies Through Integrated Performance, Process, Project, and Risk Management* (Norwood, MA: Artech House, 2008).

Tom Culhane, "Water Rights and Mitigation in Washington" (Seventh Washington Hydrogeology Symposium, Greater Tacoma Convention and Trade Center, Tacoma, WA, April 28, 2009).

Jody Freeman and Charles Kolstad, eds., *Moving to Markets in Environmental Regulation: Lessons from Twenty Years of Experience* (New York: Oxford University Press, 2006).

Norris Hundley, *The Great Thirst* (Berkeley and Los Angeles: University of California Press, 2001).

Michael Collier and Robert H. Webb, *Floods, Droughts, and Climate Change* (Tucson: University of Arizona Press, 2002).

Steven Solomon, *Water: The Epic Struggle for Wealth, Power, and Civilization* (New York: HarperCollins, 2010).

p. 192 In 2008 Harvard physician Atul Gawande published *The Checklist Manifesto*
Atul Gawande, *The Checklist Manifesto: How to Get Things Right* (New York: Metropolitan Books, 2009).

Philip K. Howard, "Problems with Protocols," *Wall Street Journal,* January 21, 2010.

p. 193 Gawande also explains how checklists can be used

Atul Gawande, *The Checklist Manifesto: How to Get Things Right* (New York: Metropolitan Books, 2009).

p. 195 One of the finest examples of the effectiveness of parallel mitigations

Howard Zinn, *A People's History of the United States, 1492–Present* (New York: Harper-Perennial, 2005).

David M. Kennedy, *The American People in World War II* (New York: Oxford University Press, 1999).

Tom Brokaw, *The Greatest Generation* (New York: Dell, 1998).

Tom Brokaw, *An Album of Memories: Personal Histories from the Greatest Generation* (New York: Random House, 2002).

James L. Stokesbury, "World War II," World Book Advanced. *World Book,* 2010. Web. July 16, 2010.

"Brief History of World War Two Advertising Campaigns War Loans and Bonds," Duke University Libraries, Digital Collections, http://library.duke.edu/digitalcollections/adaccess/warbonds.html.

p. 199 It's called venture capital

Michael Gurau, "In Baseball and Venture Capital, Success Is Batting .300," *New Hampshire Business Review,* September 25, 2007.

Interviews with David Prend, principal, Rockport Capital Partners, Palo Alto, CA, 2008–2009.

Tommi Rasila, "In Search of the Optimal Venture-to-Capital (V2C) Business Model" (Tampere, Finland: Tampere University of Technology).

John R. M. Hand, *Determinants of the Returns to Venture Capitalists* (Chapel Hill, NC: January 5, 2004).

Thomas Reuters, *2009 Venture Capital Yearbook* (Arlington, VA: National Venture Capital Association, 2009).

Geoffrey H. Smart, *What Makes a Successful Venture Capitalist?* (Chicago: Ignite Group, 2000).

Fred Wilson, "Why Early-Stage Venture Investments Fail," unionsquareventures.com, November 30, 2007, http://unionsquareventures.com/2007/11/why-early-stage.php.

Gavin C. Reid, *Risk Appraisal and Venture Capital in High Technology New Ventures* (New York: Routledge, 2007).

Michael Carusi and Prayeen Gupta, *The Ways of the VC* (Boston: Aspatore Books, 2003).

Inside CRM Editors, "Failures—Exposed, Reflected upon, Considered: The 20 Worst Venture Capital Investments of All Time," fail92fail.wordpress.com, November 8, 2008, http://fail92fail.wordpress.com/2008/11/08/the-20-worst-venture-capital-investments-of-all-time.

Ruthann Quindlen, *Confessions of a Venture Capitalist: Inside the High-Stakes World of Start-Up Financing* (New York: Warner Books, 2001).

Multiple Authors, *Green Venture Capital: Leading VCs on Analyzing Greentech Market Opportunities, Evaluating Investment Potential and Risks and Predicting the Future for Green Investing* (Boston: Aspatore Books, 2009).

p. 202 According to Nick Nuttall

Nick Nuttall, "Overfishing: A Threat to Marine Biodiversity," www.un.org/events/ten stories/06/story.asp?storyID=800.

Andrew Rosenberg, "Overfishing," *Science and Technology,* July 30, 2003.

Peter Weber, "Oceans in Peril (Overfishing and Other Problems)," *Earth Action Network*, July 28, 2005.

Carl Safina, "Where Have All the Fishes Gone?" *Science and Technology*, July 28, 2005.

"Overfishing," www.greenpeace.org.

R. Kunzig, "Twilight of the Cod," *Discover Magazine*, April 1995.

James Owen, "Overfishing Is Emptying World's Rivers, Lakes, Experts Warn," *National Geographic News*, December 1, 2005.

"Only 50 Years Left for Sea Fish," BBC News, November 2, 2006.

"Peruvian Anchovy Case: Anchovy Depletion and Trade," www.american.edu/TED/anchovy.htm.

Charles Clover, *End of the Line: How Overfishing Is Changing the World and What We Eat* (London: Ebury Press, 2004).

Suzanne Iudicello, Nuchael L. Weber, and Robert Wieland, *Fish, Markets, and Fishermen: The Economics of Overfishing* (Washington, DC: Island Press, 1999).

Richard Ellis, *The Empty Ocean* (Washington, DC: Island Press, 2003).

p. 202 Fisheries in the North Sea and Baltic Sea

"Overfishing," www.greenpeace.org.

p. 207 There was very little tangible evidence

Center for American Progress, official Web site, Iraq War Timeline: http://thinkprogress.org/iraq-timeline/.

State Department, official Web site: http://history.state.gov/departmenthistory/people/powell-colin-luther.

UN News Center. "Powell presents US case to Security Council of Iraq's failure to disarm." 5 Feb 2003. Official Web site: http://www.un.org/apps/news/story.asp?NewsID=6079&Cr=iraq&Cr1=inspect.

p. 208 Truman began discreetly inviting scientists

Interview with Charles Townes, University of California, Berkeley, Berkeley, 2009.

Charles H. Townes, *How the Laser Happened* (New York: Oxford University Press, 1999).

United States President's Science Advisory Committee Records, 1957–1961, www.aip.org/history/nbl/icos.html.

Richard L. Garwin, "Presidential Science Advising" (Cambridge, MA: Harvard University, Kennedy School of Government. Submitted for Publication in Technology and Society, September 5, 1979).

Richard Garwin, "How the Mighty Have Fallen," *Nature*, October 4, 2007.

William T. Golden, *Science Advice to the President* (Oxford: Pergamon, 1994).

David Dickson, *The New Politics of Science* (Chicago: University of Chicago Press, 1984).

Daniel S. Greenberg, John Maddox, and Steve Shapin, *The Politics of Pure Science* (Chicago: University of Chicago Press, 1999).

William T. Golden, *Science and Technology Advice to the President, Congress, and Judiciary* (Piscataway, NJ: Transaction, 1994).

Heather Douglas, *Science, Policy, and the Value-Free Ideal* (Pittsburgh: University of Pittsburgh Press, 2009).

M. Granger Morgan and John Peha, *Science and Technology Advice for Congress* (Washington, DC: RFF Press, 2003).

Benjamin P. Greene, *Eisenhower, Science Advice, and the Nuclear Test-Ban Debate 1945–1963* (Palo Alto, CA: Stanford University Press, 2006).

Chapter 11.
Bridging the Gap: Building Better Brains

p. 216 As Dr. Barry L Beyerstein, professor of psychology

Barry L. Beyerstein, "Do We Really Use Only 10 Percent of Our Brains?" *Scientific American*, March 8, 2004.

Sergio Della Sala, *Mind Myths: Exploring Popular Assumptions About the Mind and Brain* (Chichester, UK: Wiley, 1999).

Barry L. Beyerstein, *The Skeptical Inquirer*, www.csicop.org/author/barrylbeyerstein.

Daniel Druckman and John A. Swets, *Enhancing Human Performance: Issues, Theories, and Techniques* (Washington, DC: National Academy Press, 1988).

Shawn Smith, "Do We Really Only Use 10% of Our Brains?" ironshrink.com, April 16, 2007, www.ironshrink.com/articles.php?artID=070416_ten_percent_of_my_brain.

William James, *On Vital Reserves: The Energies of Men, the Gospel of Relaxation* (New York: Henry Holt, 1922)

p. 217 In the words of Dr. Elkhonon Goldberg, clinical professor of neurology

Alvaro Fernandez, "Cognitive Training and Brain Fitness Computer Programs: Interview with Dr. Elkhonon Goldberg," *Scientific American*, December 8, 2006.

Elkhonon Goldberg, *The Executive Brain: Frontal Lobes and the Civilized Mind* (New York: Oxford University Press, 2001).

p. 217 vibrantBrains was the inspiration of Jan Zivic and Lisa Schoonerman

Interview with Lisa Schoonerman, telephone, 2008.

Interview with Dr. Michael Merzenich, University of California Medical Center, Keck Center, San Francisco, 2009.

Kelly Greene, "The Latest in Mental Health: Working Out at the 'Brain Gym,'" *Wall Street Journal*, March 28, 2009.

Gordy Slack, "Brains of Steel," *San Francisco Magazine*, March 2009.

Kathleen Phalen Tomaselli, "Steps to a Nimble Mind: Physical and Mental Exercise Help Keep the Brain Fit," amednews.com, November 11, 2008, www.ama-assn.org/amednews/2008/11/17/hlsa1117.htm.

Alvaro Fernandez and Elkhonon Goldberg, *The SharpBrains Guide to Brain Fitness: 18 Interviews with Scientists, Practical Advice, and Product Reviews to Keep Your Brain Sharp* (San Francisco, CA: SharpBrains, 2009).

p. 218 Americans spent a stunning $80 million on brain mental fitness products in 2008

Kelly Greene, "The Latest in Mental Health: Working Out at the 'Brain Gym,'" *Wall Street Journal*, March 28, 2009.

p. 220 The impetus behind vibrantBrains' Neurobics Circuit is "Jewel Diver"

Interview with Dr. Michael Merzenich, University of California Medical Center, Keck Center, San Francisco, 2009.

Gordy Slack, "Brains of Steel," *San Francisco Magazine*, March 2009.

Michael Merzenich, *Studies on Functional Outcomes of Brain Fitness* (San Francisco: PositScience, 2009).

Presentation by Michael Merzenich at Oregon State University, Corvallis, May 1, 2009.

Michael Merzenich, *Brain Speed Test Results* (San Francisco: PositScience, 2009).

Katherine Ellison, "Video Games vs. the Aging Brain," *Discover Magazine*, May 21, 2007.

Kaspar Mossman, "Brain Trainers: A Workout for the Mind," *Scientific American*, April 2009.

Simon J. Evans and Paul R. Burghardt, *BrainFit for Life: A User's Guide to Life-Long Brain Health and Fitness* (Milan, MI: River Pointe, 2008).

p. 221 Merzenich, however, takes a practical view

Interview with Dr. Michael Merzenich, University of California Medical Center, Keck Center, San Francisco, 2009.

p. 221 Gordy Slack, reports on the results

Gordy Slack, "Brains of Steel," *San Francisco Magazine*, March 2009.

p. 221 Studies performed at the W. M. Keck Foundation Center for Integrative Neuroscience at UCSF

Michael Merzenich, *Studies on Functional Outcomes of Brain Fitness* (San Francisco: PositScience, 2009).

Michael Merzenich, comment on "An Insight for Successful Aging," *On the Brain*, April 25, 2008, http://merzenich.positscience.com/?p=152.

Kathleen Phalen Tomaselli, "Steps to a Nimble Mind: Physical and Mental Exercise Help Keep the Brain Fit," amednews.com, November 11, 2008, www.ama-assn.org/amednews/2008/11/17/hlsa1117.htm.

p. 222 The National Institute of Aging (NIA) began testing a "Useful Field of View"

National Institute on Aging, comp., "The Changing Brain in Health and Aging: ACTIVE Study May Provide Clues to Help Older Adults Stay Mentally Sharp," October 27, 2009, www.nia.nih.gov/Alzheimers/Publications/Unraveling/Part1/changing.htm.

Michael Merzenich, comment on "An Insight for Successful Aging," *On the Brain*, April 25, 2008, http://merzenich.positscience.com/?p=152.

Karlene Ball, "Speed Training with Older Adults: Who Benefits, for How Long, and in What Ways?" (National Institute on Aging Symposium on Cognitive Training for Older Adults, Bethesda, MD, February 29, 2004).

Michael Marsiske, "Considering the Transfer Question in Cognitive Interventions: Three Studies and Conceptual Considerations" (National Institute on Aging Symposium on Cognitive Training for Older Adults, Bethesda, MD, February 29, 2004).

Sherry Willis, "Cognitive Training on Reason Ability Within a Longitudinal Context" (National Institute on Aging Symposium on Cognitive Training for Older Adults, Bethesda, MD, February 29, 2004).

George Rebok, "Training Memory Abilities in Older Adults: In Search of Model Methods" (National Institute on Aging Symposium on Cognitive Training for Older Adults, Bethesda, MD, February 29, 2004).

John Dunlosky, "Training Metacognitive Skills to Enhance Learning" (National Institute on Aging Symposium on Cognitive Training for Older Adults, Bethesda, MD, February 29, 2004).

Wendy Rogers, "Training a System Mental Representation: Understanding Transfer of Training in the Context of Enhanced Activities of Daily Living" (National Institute on Aging Symposium on Cognitive Training for Older Adults, Bethesda, MD, February 29, 2004).

Sara Czaja, "Training and the Acquisition of Real World Functional Tasks" (National Institute on Aging Symposium on Cognitive Training for Older Adults, Bethesda, MD, February 29, 2004).

Kathleen Phalen Tomaselli, "Steps to a Nimble Mind: Physical and Mental Exercise Help Keep the Brain Fit," amednews.com, November 11, 2008, www.ama-assn.org/amednews/2008/11/17/hlsa1117.htm.

Daniel Druckman and John A. Swets, *Enhancing Human Performance: Issues, Theories, and Techniques* (Washington, DC: National Academy Press, 1988).

Alvaro Fernandez, "Cognitive Training and Brain Fitness Computer Programs: Interview with Dr. Elkhonon Goldberg," *Scientific American*, December 8, 2006.

National Institute on Aging, comp., "The Changing Brain in Health and Aging: ACTIVE Study May Provide Clues to Help Older Adults Stay Mentally Sharp," October 27, 2009, www.nia.nih.gov/Alzheimers/Publications/Unraveling/Part1/changing.htm.

Jake Dunagan, "Pumping Up the Brain. Reflections on the SharpBrains Virtual Summit," *Scientific American*, February 9, 2010.

Lawrence J. Whalley, *The Aging Brain* (New York: Columbia University Press, 2003).

Patrick R. Hop and Charles V. Mobbs, *Functional Neurobiology of Aging* (San Diego: Academic Press, 2000).

William Jagust and Mark D'esposito, *Imaging the Aging Brain* (New York: Oxford University Press, 2009).

p. 222 The *Journal of American Geriatrics Society* also confirmed the findings

Kelly Greene, "The Latest in Mental Health: Working Out at the 'Brain Gym,'" *Wall Street Journal*, March 28, 2009.

Richard Restak, *The New Brain: How the Modern Age Is Rewiring Your Mind* (Emmaus, PA: Rodale, 2004).

National Institute on Aging, comp., "The Changing Brain in Health and Aging: ACTIVE Study May Provide Clues to Help Older Adults Stay Mentally Sharp," October 27, 2009, www.nia.nih.gov/Alzheimers/Publications/Unraveling/Part1/changing.htm.

Michael Merzenich, *Studies on Functional Outcomes of Brain Fitness* (San Francisco: PositScience, 2009).

Michael Merzenich, comment on "An Insight for Successful Aging," *On the Brain*, April 25, 2008, http://merzenich.positscience.com/?p=152.

Presentation by Michael Merzenich at Oregon State University, Corvallis, May 1, 2009.

p. 223 Merzenich is best known for his pioneering work on a phenomenon called brain "plasticity"

Michael Merzenich, *On the Brain: About Brain Plasticity* (San Francisco: PositScience, April 16, 2008).

Michael Merzenich, "On Rewiring the Brain" (speech from the TED Conference, Monterey, CA, February 26, 2004).

Erin Clifford, "Neural Plasticity: Merzenich, Taub, and Grenough," *The Harvard Brain* 6(1) (1999).

J. E. Black and W. T. Grenough, *Neurobiology of Learning and Memory* (San Diego: Academic Press, 1998).

D. V. Buonomano and Michael Merzenich, "Cortical Plasticity: From Synapses to Maps," *Annual Review of Neuroscience* 21 (March 1998).

M. Merzenich, J. H. Kaas, J. Wall, R. J. Nelson, M. Sur, and D. Felleman, "Topographic Reorganization of Somasensory Cortical Areas 3b and I in Adult Monkeys Following Restricted Deafferentation," *Neuroscience* 8(1) (January 1983).

J. P. Raushecker, "Compensatory Plasticity and Sensory Substitution in the Cerebral Cortex," *Trends in Neuroscience* 18(1) (1995).

G. H. Recanzone, C. E. Schreiner, and Michael Merzenich, "Plasticity in the Frequency Representation of Primary Auditory Cortex Following Training in Adult Owl Monkeys," *Journal of Neuroscience* 13(1) (1993).

"Harnessing the Brain's Plasticity Key to Treating Neurological Damage," *Science Daily*, February 27, 2007.

Ginger Campbell, "Michael Merzenich on Neuroplasticity," Brain Science podcast, Episode 54, February 13, 2009.

Wotjek Chodzko-Zajko, Arthur Kramer, and Leonard Poon, eds., *Enhancing Cognitive Functioning and Brain Plasticity* (Champaign, IL: Human Kinetics, 2009).

Brian Kolb, *Brain Plasticity and Behavior* (Philadelphia: Psychology Press, 1995).

Guido Filogamo, Antonia Vernadakis, Fulvia Gremo, and Alain M. Privat, *Brain Plasticity: Development and Aging* (New York: Plenum Press, 1997).

Kathleen Phalen Tomaselli, "Steps to a Nimble Mind: Physical and Mental Exercise Help Keep the Brain Fit," amednews.com, November 11, 2008, www.ama-assn.org/amednews/2008/11/17/hlsa1117.htm.

Barbara Strauch, "How to Train the Aging Brain," *New York Times*, December 29, 2009.

Norman Doidge, *The Brain That Changes Itself: Stories of Personal Triumph from the Frontiers of Brain Science* (New York: Penguin, 2007).

p. 223   Merzenich provides an easy example of how the human brain can learn
Presentation by Michael Merzenich at Oregon State University, Corvallis, May 1, 2009.

p. 224   How the brain prioritizes what it wants to learn
Gordy Slack, "Brains of Steel," *San Francisco Magazine*, March 2009.

p. 224   "Neurons that fire together wire together"
Gordy Slack, "Brains of Steel," *San Francisco Magazine*, March 2009.

p. 225   "Chance favors the *prepared* mind"
J. Kounios, J. L. Frymiare, E. M. Bowden, J. I. Fleck, K. Subramaniam, T. B. Parnish, and M. Jung-Beeman, "The Prepared Mind: Neural Activity Prior to Problem Presentation Predicts Subsequent Solution by Sudden Insight," *Psychological Sciences* 17(10) (2006).

p. 225   Schoolchildren in the United States are already using brain fitness technology
Presentation by Michael Merzenich at Oregon State University, Corvallis, May 1, 2009.

Interview with Dr. Michael Merzenich, University of California Medical Center, Keck Center, San Francisco, 2009.

Michael Merzenich, *Studies on Functional Outcomes of Brain Fitness* (San Francisco: PositScience, 2009).

Ginger Campbell, "Michael Merzenich on Neuroplasticity," Brain Science podcast, Episode 54, February 13, 2009.

Michael Merzenich, "On Rewiring the Brain" (speech from the TED Conference, Monterey, CA, February 26, 2004).

Various case studies, Scientific Learning Corporation, www.scilearn.com.

"Cumberland County School District See 'Amazing Results' for All Types of Students Using Fast Forward Software," Scientific Learning Corporation, www.scilearn.com/alldocs/mktg/10214CumberlandCS.pdf.

Judy Willis, *Research-Based Strategies to Ignite Student Learning: Insights from a Neurologist and Classroom Teacher* (Alexandria, VA: Association for Supervision and Curriculum Development, August 30, 2006).

Antonio M. Bahiro, Kurt W. Fischer, and Pierre J. Léna, *The Educated Brain: Essays in Neuroeducation* (New York: Cambridge University Press, 2008).

M. Layne Kalbfleisch, "Getting to the Heart of the Brain: Using Cognitive Neuroscience to Explore the Nature of Human Ability and Performance," *Roeper Review*, December 2, 2008.

John Geake, *The Brain at School: Educational Neuroscience in the Classroom* (Maidenhead, UK: Open University Press, 2009).

p. 226 Children who were exposed to brain fitness showed *twice* the academic achievement

Presentation by Michael Merzenich at Oregon State University, Corvallis, May 1, 2009.

Interview with Dr. Michael Merzenich, University of California Medical Center, Keck Center, San Francisco, 2009.

Mary Ann Petrillo, "Jackson County (MS) School District Selected as National Reference Site," News release from Scientific Learning Corporation, February 2, 2007, http://www.scilearn.com/company/news/news-releases/20070208.php.

p. 227 The answer lies in "spontaneous thought"

Kalina Christoff, Alan Gordon, and Rachelle Smith, "The Role of Spontaneous Thought in Human Cognition," in *Neuroscience of Decision Making*, ed. Oshin Vartanian and David R. Mandel (New York: Psychology Press, n.d.).

A. M. Achim, M. Bertrand, A. Montoya, A. K. Malla, and M. Lepage, "Medial Temporal Lobe Activations During Associative Memory Encoding for Arbitrary and Semantically Related Object Pairs," *Brain Research* 1161(3) (August 2007).

N. C. Andreasen, D. S. O'Leary, T. Cizadlo, S. Arnot, K. Rezai, G. L. Watkins, L. L. Ponto, and R. D. Hichwa, "Remembering the Past: Two Facets of Episodic Memory Explored with Positron Emission Tomography," *American Journal of Psychiatry* 152(11) (1995).

p. 228 People who have insights report a sudden, unexpected ability to "see connections"

Hemai Parthasarathy, "Imagining the Brain: Solving Problems Through Insight," *PLoS Biology* 1(4) (April 13, 2004).

E. M. Bowden, M. Jung-Beeman, J. Fleck, and J. Jounios, "New Approaches to Demystifying Insight," *Trends in Cognitive Sciences* 9(7) (2005).

Dianna Amorde and Christine Frank, *Aha! Moments: When Intellect and Intuition Collide* (Boston: Inspired Press, 2009).

A. J. K. Pols, "Insight in Problem Solving," http://www.phil.uu.nl/preprints/ckiscripties/SCRIPTIES/018_pols.pdf.

Jing Luo and Guenther Knoblich, "Studying Insight Problem Solving with Neuroscientific Methods," *ScienceDirect*, December 7, 2006.

Jonah Lehrer, "The Eureka Hunt," *New Yorker*, July 28, 2008.

"Brain Activity Differs for Creative and Noncreative Thinkers," *Science Daily*, October 29, 2007.

"Neural Basis of Solving Problems with Insight," *PLoS Biology* 2(4) (2004).

David Rock and Jeffrey Schwartz, "The Neuroscience of Leadership," *Strategy + Business* 43 (2006).

Jeffrey Schwartz and Sharon Begley, *The Mind and the Brain: Neuroplasticity and the Power of Mental Force* (New York: ReganBooks, 2002).

Jeffrey Schwartz, Henry P. Stapp, and Mario Beauregard, "Quantum Physics in Neuroscience and Psychology: A Neurophysical Model of the Mind-Brain Interaction," *Proceedings of the Royal Society B: Biological Sciences* 360(1458) (June 29, 2005).

Kalina Christoff, Justin M. Ream, and John D. E. Gabieli, "Neural Basis of Spontaneous Thought Processes," *Cortex* 40(4) (2004).

J. Kounios, J. L. Frymiare, E. M. Bowden, J. I. Fleck, K. Subramaniam, T. B. Parnish, and M. Jung-Beeman, "The Prepared Mind: Neural Activity Prior to Problem Presentation Predicts Subsequent Solution by Sudden Insight," *Psychological Sciences* 17(10) (2006).

A. Dijksterhuis, "A Theory of Unconscious Thought," *Perspectives on Psychological Science* 1(2) (2006).

Jennifer Dorfman, Victor A. Shames, and John F. Kihlstrom, "Intuition, Incubation, and Insight: Implicit Cognition in Problem Solving," in *Implicit Cognition*, ed. Geoffrey Underwood (New York: Oxford University Press, 1996).

Jonathan W. Schooler and Joseph Melcher, "The Ineffability of Insight," in *The Creative Cognition Approach*, ed. Steven M. Smith, Thomas B. Ward, and Ronald A. Fink (Cambridge, MA: MIT Press, 1995).

John Kounios, Jessica I. Fleck, Deborah L. Green, Lisa Payne, Jennifer L. Stevenson, Edward M. Bowden, and Mark Jung-Beeman, "The Origins of Insight in Resting-State Brain Activity" (Philadelphia: Department of Psychology, Drexel University, July 18, 2007).

Yun Chu, *Human Insight Problem Solving: Performance, Processing, and Phenomenology* (Saarbrücken, Germany: VDM Verlag, 2009).

Martin Gardner, *Aha! Insight* (New York: Freeman, 1978).

Robert J. Sternberg and Janet E. Davidson, *The Nature of Insight* (Cambridge, MA: MIT Press, 1995).

Robert J. Sternberg and Talia Ben-Zeev, *Complexity Cognition: The Psychology of Human Thought* (New York: Oxford University Press, 2001).

P. I. Ansburg, *Current Psychology*, 2000.

P. I. Ansburg and R. L. Dominowski, *Journal of Creative Behavior*, 2000.

Baker-Sennett and S. J. Ceci, *Journal of Creative Behavior*, 1996.

E. M. Bowden and M. J. Beeman, *Psychological Science*, 1998.

E. P. Chronicle, Y. C. Omerod, and J. N. MacGregor, *Quarterly Journal of Experimental Psychology Section A—Human Experimental Psychology*, 2001.

Robert J. Sternberg and Janet E. Davidson, *The Nature of Insight* (Cambridge, MA: MIT Press, 1995).

C. A. Kaplan and H. A. Simon, "In Search of Insight," *Cognitive Psychology* 22(3) (July 1990).

G. Knoblich, S. Ohlsson, H. Haider, and D. Rhebius, *Journal of Experimental Psychology: Learning, Memory, and Cognition*, 1999.

R. S. Lockhart, M. Lamon, and M. L. Gick, *Memory and Cognition*, 1988.

N. R. F. Maier, "Reasoning in Humans II: The Solution of a Problem and Its Appearance in Consciousness," *Journal of Comparative Psychology* 13 (1931).

Richard E. Mayer, *Thinking, Problem Solving, and Cognition* (New York: Freeman, 1992).

J. Metcalfe, *Journal of Experimental Psychology: Learning, Memory, and Cognition*, 1986.

J. Metcalfe and D. Wiebe, *Memory and Cognition*, 1987.

N. Mori, *Japanese Psychological Research*, 1996.

A. Del Cul, S. Baillet, and S. Dehaene, *PLoS Biology*, 2007.

P. Haggard and B. Libel, *Journal of Consciousness Studies*, 2001.

Sandkuhler S. Bhattacharya, "Deconstructing Insight: EEG Correlates of Insightful Problem Solving," *PLoS ONE* 3(1) (2008).

Jonathan W. Schooler and Joseph Melcher, "The Ineffability of Insight" (paper presented at the annual meeting of the Psychonomic Society, Washington, DC, 1993).

R. S. Siegler, "Unconscious Insights," *Current Directions in Psychological Science* 9(3) (2000).

R. W. Smith and J. Kounios, "Sudden Insight," *Journal of Experimental Psychology: Learning, Memory, and Cognition* 22(6) (1996).

R. W. Weisberg, *Journal of Experimental Psychology: Learning, Memory, and Cognition,* 1992.

M. J. Beeman and E. M. Bowden, *Memory and Cognition,* 2000.

E. M. Bowden and M. J. Beeman, *Psychological Science,* 1998.

E. M. Bowden and M. Jung-Beeman, "Aha! Insight Experience Correlates with Solution Activation in the Right Hemisphere," *Psychonomic Bulletin and Review* 10(3) (2003).

E. M. Bowden, M. Jung-Beeman, J. Fleck, and J. Jounios, "New Approaches to Demystifying Insight," *Trends in Cognitive Sciences* 9(7) (2005).

Nancy C. Andreasen, *The Creating Brain: The Neuroscience of Genius* (Washington, DC: Dana Press, 2005).

E. Angelakis, J. F. Lubar, S. Stathopoulou, and J. Kounios, "Peak Alpha Frequency: An Electroencephalographic Measure of Cognitive Preparedness," *Clinical Neurophysiology* 115(4) (2004).

M. Jung-Beeman, E. M. Bowden, J. Haberman, J. L. Frymiare, S. Arambel-Liu, and R. Greenblatt, "Neural Activity When People Solve Problems with Insight," *PLoS Biology,* 2004.

J. Kounios, A. M. Osman, and D. E. Meyer, "Structure and Process in Semantic Memory: New Evidence Based on Speed-Accuracy Decomposition," *Journal of Experimental Psychology* 90(1–3) (1987).

S. Lang, N. Kanngieger, P. Jaskowski, H. Haider, M. Rose, and R. Verleger, "Precursors of Insight in Event-Related Brain Potentials," *Journal of Cognitive Neuroscience* 18(12) (2006).

A. Newell and H. A. Simon, *Human Problem Solving* (Englewood Cliffs, NJ: Prentice Hall, 1972).

K. Christoff and J. D. E. Gabrieli, "The Frontopolar Cortex and Human Cognition," *Psychobiology* 28(2) (2000).

Geoffrey Underwood, ed., *Implicit Cognition* (New York: Oxford University Press, 1996).

Z. Chen and Daehler, "External and Internal Instantiation of Abstract Information Facilitates Transfer in Insight Problem Solving," *Contemporary Educational Psychology* 25(4) (2000).

T. C. Kershaw and S. Ohlsson, "Training for Insight: The Case of the Nine-Dot Problem" (paper presented at the Twenty-third Annual Meeting of the Cognitive Science Society, Edinburgh, Scotland, August 1–4, 2001).

M. Weith and B. D. Burns, "Motivation in Insight Versus Incremental Problem Solving" (paper presented at the Twenty-second Meeting of the Cognitive Science Society, University of Pennsylvania, Philadelphia, PA, August 13–15, 2000).

p. 228 According to Dr. Karuna Subramaniam

Karuna Subramaniam, John Kounios, B. Todd, and Mark Jung-Beeman, "A Brain Mechanism for Facilitation of Insight by Positive Affect," *Journal of Cognitive Neuroscience* 21(3) (2009).

p. 229 In the 2006 article "The Prepared Mind"

J. Kounios, J. L. Frymiare, E. M. Bowden, J. I. Fleck, K. Subramaniam, T. B. Parnish, and M. Jung-Beeman, "The Prepared Mind: Neural Activity Prior to Problem Presentation Predicts Subsequent Solution by Sudden Insight," *Psychological Sciences* 17(10) (2006).

A. J. K. Pols, "Insight in Problem Solving," http://www.phil.uu.nl/preprints/ckiscripties/SCRIPTIES/018_pols.pdf.

Jing Luo and Guenther Knoblich, "Studying Insight Problem Solving with Neuroscientific Methods," *ScienceDirect*, December 7, 2006.

p. 230 Insight is extremely cognitively taxing

J. Kounios, J. L. Frymiare, E. M. Bowden, J. I. Fleck, K. Subramaniam, T. B. Parnish, and M. Jung-Beeman, "The Prepared Mind: Neural Activity Prior to Problem Presentation Predicts Subsequent Solution by Sudden Insight," *Psychological Sciences* 17(10) (2006).

G. Dreisbach and T. Goschke, "How PA Modulates Cognitive Control: Reduced Perseveration at the Cost of Increased Distractability," *Journal of Experimental Psychology: Learning, Memory, and Cognition* 30(2) (2004).

Karuna Subramaniam, John Kounios, B. Todd, and Mark Jung-Beeman, "A Brain Mechanism for Facilitation of Insight by Positive Affect," *Journal of Cognitive Neuroscience* 21(3) (2009).

Kalina Christoff, Alan Gordon, and Rachelle Smith, "The Role of Spontaneous Thought in Human Cognition," in *Neuroscience of Decision Making*, ed. Oshin Vartanian and David R. Mandel (New York: Psychology Press, n.d.).

John Kounios, Jessica I. Fleck, Deborah L. Green, Lisa Payne, Jennifer L. Stevenson, Edward M. Bowden, and Mark Jung-Beeman, "The Origins of Insight in Resting-State Brain Activity" (Philadelphia: Department of Psychology, Drexel University, July 18, 2007).

Hemai Parthasarathy, "Imagining the Brain: Solving Problems Through Insight," *PLoS Biology*, April 13, 2004.

E. M. Bowden, M. Jung-Beeman, J. Fleck, and J. Jounios, "New Approaches to Demystifying Insight," *Trends in Cognitive Sciences* 9(7) (2005).

Dianna Amorde and Christine Frank, *Aha! Moments: When Intellect and Intuition Collide* (Boston: Inspired Press, 2009).

A. J. K. Pols, "Insight in Problem Solving," http://www.phil.uu.nl/preprints/ckiscripties/SCRIPTIES/018_pols.pdf.

Jing Luo and Guenther Knoblich, "Studying Insight Problem Solving with Neuroscientific Methods," *ScienceDirect*, December 7, 2006.

Jonah Lehrer, "The Eureka Hunt," *New Yorker*, July 28, 2008.

"Brain Activity Differs for Creative and Noncreative Thinkers," *Science Daily*, October 29, 2007.

"Neural Basis of Solving Problems with Insight," *PLoS Biology* 2(4) (2004).

David Rock and Jeffrey Schwartz, "The Neuroscience of Leadership," *Strategy + Business* 43 (2006).

Jeffrey Schwartz and Sharon Begley, *The Mind and the Brain: Neuroplasticity and the Power of Mental Force* (New York: ReganBooks, 2002).

Jeffrey Schwartz, Henry P. Stapp, and Mario Beauregard, "Quantum Physics in Neuroscience and Psychology: A Neurophysical Model of the Mind-Brain Interaction," *Proceedings of the Royal Society B: Biological Sciences* 360(1458) (June 29, 2005).

Kalina Christoff, Justin M. Ream, and John D. E. Gabieli, "Neural Basis of Spontaneous Thought Processes," *Cortex* 40(4) (2004).

A. Dijksterhuis, "A Theory of Unconscious Thought," *Perspectives on Psychological Science* 1(2) (2006).

Jennifer Dorfman, Victor A. Shames, and John F. Kihlstrom, "Intuition, Incubation, and Insight: Implicit Cognition in Problem Solving," in *Implicit Cognition*, ed. Geoffrey Underwood (New York: Oxford University Press, 1996).

Jonathan W. Schooler and Joseph Melcher, "The Ineffability of Insight," in *The Creative Cognition Approach*, ed. Steven M. Smith, Thomas B. Ward, and Ronald A. Fink (Cambridge, MA: MIT Press, 1995).

Yun Chu, *Human Insight Problem Solving: Performance, Processing, and Phenomenology* (Saarbrücken, Germany: VDM Verlag, 2009).

Martin Gardner, *Aha! Insight* (New York: Freeman, 1978).

Robert J. Sternberg and Janet E. Davidson, *The Nature of Insight* (Cambridge, MA: MIT Press, 1995).

Robert J. Sternberg and Talia Ben-Zeev, *Complexity Cognition: The Psychology of Human Thought* (New York: Oxford University Press, 2001).

P. I. Ansburg, *Current Psychology*, 2000.

P. I. Ansburg and R. L. Dominowski, *Journal of Creative Behavior*, 2000.

Baker-Sennett and S. J. Ceci, *Journal of Creative Behavior*, 1996.

E. M. Bowden and M. J. Beeman, *Psychological Science*, 1998.

E. P. Chronicle, Y. C. Omerod, and J. N. MacGregor, *Quarterly Journal of Experimental Psychology Section A—Human Experimental Psychology*, 2001.

Robert J. Sternberg and Janet E. Davidson, *The Nature of Insight* (Cambridge, MA: MIT Press, 1995).

C. A. Kaplan and H. A. Simon, "In Search of Insight," *Cognitive Psychology* 22(3) (July 1990).

G. Knoblich, S. Ohlsson, H. Haider, and D. Rhebius, *Journal of Experimental Psychology: Learning, Memory, and Cognition*, 1999.

R. S. Lockhart, M. Lamon, and M. L. Gick, *Memory and Cognition*, 1988.

N. R. F. Maier, "Reasoning in Humans II: The Solution of a Problem and Its Appearance in Consciousness," *Journal of Comparative Psychology* 13 (1931).

Richard E. Mayer, *Thinking, Problem Solving, and Cognition* (New York: Freeman, 1992).

J. Metcalfe, *Journal of Experimental Psychology: Learning, Memory, and Cognition*, 1986.

J. Metcalfe and D. Wiebe, *Memory and Cognition*, 1987.

N. Mori, *Japanese Psychological Research*, 1996.

P. Haggard and B. Libel, *Journal of Consciousness Studies*, 2001.

Sandkuhler S. Bhattacharya, "Deconstructing Insight: EEG Correlates of Insightful Problem Solving," *PLoS ONE* 3(1) (2008).

Jonathan W. Schooler and Joseph Melcher, "The Ineffability of Insight" (paper presented at the annual meeting of the Psychonomic Society, Washington, DC, 1993).

R. S. Siegler, "Unconscious Insights," *Current Directions in Psychological Science* 9(3) (2000).

R. W. Smith and J. Kounios, "Sudden Insight," *Journal of Experimental Psychology: Learning, Memory, and Cognition* 22(6) (1996).

R. W. Weisberg, *Journal of Experimental Psychology: Learning, Memory, and Cognition*, 1992.

M. J. Beeman and E. M. Bowden, *Memory and Cognition*, 2000.

E. M. Bowden and M. J. Beeman, *Psychological Science*, 1998.

E. M. Bowden and M. Jung-Beeman, "Aha! Insight Experience Correlates with Solution Activation in the Right Hemisphere," *Psychonomic Bulletin and Review* 10(3) (2003).

E. M. Bowden, M. Jung-Beeman, J. Fleck, and J. Jounios, "New Approaches to Demystifying Insight," *Trends in Cognitive Sciences* 9(7) (2005).

Nancy C. Andreasen, *The Creating Brain: The Neuroscience of Genius* (Washington, DC: Dana Press, 2005).

E. Angelakis, J. F. Lubar, S. Stathopoulou, and J. Kounios, "Peak Alpha Frequency: An Electroencephalographic Measure of Cognitive Preparedness," *Clinical Neurophysiology* 115(4) (2004).

M. Jung-Beeman, E. M. Bowden, J. Haberman, J. L. Frymiare, S. Arambel-Liu, and R. Greenblatt, "Neural Activity When People Solve Problems with Insight," *PLoS Biology*, 2004.

J. Kounios, A. M. Osman, and D. E. Meyer, "Structure and Process in Semantic Memory: New Evidence Based on Speed-Accuracy Decomposition," *Journal of Experimental Psychology* 90(1–3) (1987).

S. Lang, N. Kanngieger, P. Jaskowski, H. Haider, M. Rose, and R. Verleger, "Precursors of Insight in Event-Related Brain Potentials," *Journal of Cognitive Neuroscience* 18(12) (2006).

A. Newell and H. A. Simon, *Human Problem Solving* (Englewood Cliffs, NJ: Prentice Hall, 1972).

K. Christoff and J. D. E. Gabrieli, "The Frontopolar Cortex and Human Cognition," *Psychobiology* 28(2) (2000).

Geoffrey Underwood, ed., *Implicit Cognition* (New York: Oxford University Press, 1996).

p. 230 Before insight was deployed, the brain could be observed preparing itself
J. Kounios, J. L. Frymiare, E. M. Bowden, J. I. Fleck, K. Subramaniam, T. B. Parnish, and M. Jung-Beeman, "The Prepared Mind: Neural Activity Prior to Problem Presentation Predicts Subsequent Solution by Sudden Insight," *Psychological Sciences* 17(10) (2006).

p. 232 In 2004, the University of Chicago Howard Hughes Medical Institute published an article
"Human Brain Evolution Was a 'Special Event,'" *HHMI Research News*, December 29, 2004, http://www.hhmi.org/news/pdf/lahn3.pdf.

Richard Restak, *The New Brain: How the Modern Age Is Rewiring Your Mind* (Emmaus, PA: Rodale, 2004).

Nicholas Wade, "Researchers Say Human Brain Is Still Evolving," *New York Times*, September 8, 2005.

William H. Calvin, Terrence Deacon, Ralph L. Holloway, Richard G. Klein, Steven Pinker, John Tooby, Endel Tulving, and Ajit Varki, "The Evolution of the Human Brain" (Center for Human Evolution, proceedings of Workshop 5, Bellevue, WA, March 19–20, 2005).

Jane Bradbury, "Molecular Insights into Human Brain Evolution," *PLoS Biology* 3(3) (2005).

Kate Melville, "Evolution of the Human Brain Unique," Scienceagogo, December 29, 2004, http://www.scienceagogo.com/news/20041129182724data_trunc_sys.shtml.

George F. Striedter, *Principles of Brain Evolution* (Sunderland, MA: Sinauer Associates, 2004).

Christopher Willis, *The Runaway Brain: The Evolution of Human Uniqueness* (New York: Basic Books, 1994).

Richard Dawkins, *The Selfish Gene* (New York: Oxford University Press, 2006).

John R. Krebs and Richard Dawkins, *Animal Signals: Mind Reading and Manipulation*, in *Behavioral Ecology: An Evolutionary Approach*, ed. John R. Krebs and Nicholas B. Davies, 2nd ed. (Oxford: Blackwell, 1984).

Kenneth E. Boulding, *Evolutionary Economics* (Thousand Oaks, CA: Sage, 1981).

Robert H. Frank, "If 'Homoeconomicus' Could Choose His Own Utility Function, Would He Want One with a Conscience?" *American Economic Review* 79 (June 1989).

Kurt Dopfer, ed., *The Evolutionary Foundation of Economics* (Cambridge: Cambridge University Press, 2006).

Michael Shermer, *The Mind of the Market: How Biology and Psychology Shape Our Economic Lives* (New York: Holt Paperbacks, 2006).

J. Stanley Metcalfe, *Evolutionary Economics and Creative Destruction* (New York: Routledge, 1998).

Arthur Gandolfi, Anna Gandolfi, and David Barash, *Economics as an Evolutionary Science: From Utility to Fitness* (Edison, NJ: Transaction, 2002).

Daniel Friedman, *Morals and Markets: An Evolutionary Account of the Modern World* (New York: Palgave Macmillan, 2008).

Jason Potts, *The New Evolutionary Microeconomics: Complexity, Competence, and Adaptive Behavior* (Cheltenham, UK: Edward Elgar, 2001).

Peter Koslowski, *Sociobiology and Bioeconomics: The Theory of Evolution in Biological and Economic Theory* (New York: Springer, 1999).

Sunny Y. Auyand, *Foundations of Complex-System Theories in Economies, Evolutionary Biology, and Statistical Physics* (Cambridge: Cambridge University Press, 1999).

Charles Darwin, *The Descent of Man* (New York: Penguin, 2007).

Charles Darwin, *The Expression of the Emotions in Man and Animals* (New York: Oxford University Press, 1998).

Chris Stringer and Peter Andrews, *The Complete World of Human Evolution* (New York: Thames and Hudson, 2005).

Gregory Cochran and Henry Harpending, *The 10,000 Year Explosion: How Civilization Accelerated Human Evolution* (New York: Basic Books, 2009).

John Cartwright, *Evolution and Human Behavior: Darwinian Perspectives on Human Nature* (Cambridge, MA: MIT Press, 2000).

Eva Jablonka and Marion J. Lamb, *Evolution in Four Dimensions: Genetic, Epigenetic, Behavioral, and Symbolic Variation in the History of Life* (Cambridge, MA: MIT Press, 2005).

Michael J. Behe, *The Edge of Evolution: The Search for the Limits of Darwinism* (New York: Free Press, 2008).

Christopher Scarre, *The Human Past: World Prehistory and the Development of Human Societies* (New York: Thames and Hudson, 2005).

Timothy Goldsmith and William F. Zimmerman, *Biology, Evolution, and Human Nature* (New York: Wiley, 2000).

Philip Clayton and Jeffrey Schloss, *Evolution and Ethics: Human Morality in Biological and Religious Perspective* (Grand Rapids, MI: Eerdmans, 2004).

Bruce H. Lipton and Steve Bhaerman, *Spontaneous Evolution: Our Positive Future (and a Way to Get There from Here)* (Carlsbad, CA: Hay House, 2009).

Richard Restak, *The New Brain: How the Modern Age Is Rewiring Your Mind* (Emmaus, PA: Rodale, 2004).

Mihail C. Roco and Carlo D. Montemagno, *The Coevolution of Human Potential and Converging Technologies* (New York: New York Academy of Science, 2004).

Robert J. Sternberg and Janet E. Davidson, *The Nature of Insight* (Cambridge, MA: MIT Press, 1995).

p. 232 The more complex our behavior becomes, the more impact that complexity has

Serendip, comp., "Brain Size and Evolution—Where We Are . . . So as to See Where We Might Go Next," http://serendip.brynmawr.edu/bb/brain evolution.

p. 234 The connection between insight and the unconscious mind

Interview with Jerry Lauch, Louisville Sleep Disorders Center, Carmel, CA, 2009.

"Got a Problem? Think About It Overnight," *US News and World Report,* June 9, 2009.

A. Dijksterhuis and T. Meurs, "Where Creativity Resides: The Generative Powers of Unconscious Thought," *Consciousness and Cognition* 15(1) (March 2006).

A. Dijksterhuis, "A Theory of Unconscious Thought," *Perspectives on Psychological Science* 1(2) (2006).

Howard Shevrin, James A. Bond, Linda A. Grakel, Richard K. Hertel, and William Williams, *Conscious and Unconscious Processes: Psychodynamic, Cognitive, and Neurophysiological Convergences* (New York: Guilford Press, 1996).

Dan J. Stein, *Cognitive Science and the Unconscious* (Washington, DC: American Psychiatric Press, 1997).

Gerd Gigerenzer, *Gut Feelings: The Intelligence of the Unconscious,* narr. Dick Hill (Old Saybrook, CT: Tantor Media Audio Books, 2007).

Geoffrey Underwood, ed., *Implicit Cognition* (New York: Oxford University Press, 1996).

Ernest Hartman, *Dreams and Nightmares: The New Theory on the Origin and Meaning of Dreams* (New York: Plenum Press, 1998).

p. 235 Its exploration began with psychologists William James and Boris Sidis

Boris Sidis and William James, *The Psychology of Suggestion: A Research into the Subconscious Nature of Man and Society* (Whitefish, MT: Kessinger, 1896).

Sigmund Freud, *The Interpretation of Dreams* (New York: Random House, 1950).

Liliane Frey-Rohn, *From Freud to Jung: A Comparative Study of the Psychology of the Unconscious* (New York: C. G. Jung Foundation Books, 2001).

p. 236 In the spring of 2008, a dry lightning storm sparked a fire

Interview with John Saar, John Saar Real Estate, Carmel, CA, 2008.

"Big Sur Fire Rages On," KVSP.org/fire/sur.html, July 1, 2008.

"California's Continuing Fires," Boston.com/big picture, July 7, 2008, www.boston.com/bigpicture/2008/07/californias_continuing_fires.html.

p. 239 According to science writer Jonah Lehrer

Jonah Lehrer, "The Eureka Hunt," *New Yorker,* July 28, 2008.

p. 239 William Futrell, a pioneer in educational reform

Correspondence with William A. Futrell, June 4, 2009.

Chapter 12.
Invoking Insight: Conditions Conducive to Cognition

p. 244 Science writer Robert Roy Britt reports that even the most dedicated and gifted among us are getting bogged down

Robert Roy Britt, "Is Einstein the Last Great Genius?" LiveScience.com, December 5, 2008, www.livescience.com/culture/081205-science-genius-einstein.html.

John Horgan, *The End of Science: Facing the Limits of Knowledge in the Twilight of the Scientific Age* (New York: Broadway Books, 1997).

p. 244 Groups of three, four, and five people solve highly complex problems better

Patrick R. Laughlin, Erin C. Hatch, Jonathan S. Silver, and Lee Both, "Groups Perform

Better Than the Best Individuals on Letters and Numbers Problems: Effect of Group Size," *Journal of Personality and Social Psychology* 90(4) (2006).

Nancy K. Napier, *Insight: Encouraging Aha! Moments for Organizational Success* (Westport, CT: Greenwood, 2010).

Jon R. Katzenbach and Douglas K. Smith, *The Discipline of Teams: A Mindbook-Workbook for Delivering Small Group Performance* (New York: Wiley, 2001).

James R. Larson Jr., *In Search of Synergy in Small Group Performance* (London: Psychology Press, 2009).

Charles Day, "Right-Sizing the Table," *Association Management*, July 28, 2005.

Earl A. Alluisi, *The Measurement of Small Group Performance in a Restrictive Environment* (Lockheed, GA: Human Factors Research Department, 1962).

Francis L. Ulschak, *Small Group Problem Solving: An Aid to Organizational Effectiveness* (Reading, MA: Addison-Wesley, 1981).

G. W. Hill, "Group Versus Individual Performance: Are N + 1 Heads Better Than One?" *Psychological Bulletin* 91(3) (1982).

D. W. Taylor and W. L. Faust, "Twenty Questions: Efficiency in Problem-Solving as a Function of Group Size," *Journal of Experimental Psychology* 44 (1952).

E. J. Thomas and C. F. Fink, "Models of Group Problem Solving," *Journal of Abnormal and Social Psychology* 63(1) (July 1961).

p. 244 The ideal group size has been around ten individuals

Antony Jay, *Corporation Man* (New York: Pocket Books, 1973).

p. 245 The minimum group size required to efficiently solve tough problems was three persons

J. Dan Rothwell, *In Mixed Company: Small Group Communications*, 3rd ed. (New York: Harcourt Brace College, 1998).

Manny Robertson, "Group Decision-Making Process," University of Kentucky, www.uky.edu.

Nathan Zook, "If You're Working in a Big Group, You're Fighting Human Nature," April 24, 2008, http://37signals.com/svn/posts/995-if-youre-working-in-a-big-group-youre-fighting-human-nature.

Eugene M. Lewitt and Linda Schuurman Baker, "Class Size," *The Future of Children: Financing Schools* 7(3) (Winter 1997).

"Class Size Reduction: A Proven Reform Strategy" (Washington, DC: National Education Association Education Policy and Practice Department, 2009).

Patrick R. Laughlin, Erin C. Hatch, Jonathan S. Silver, and Lee Both, "Groups Perform Better Than the Best Individuals on Letters and Numbers Problems: Effect of Group Size," *Journal of Personality and Social Psychology* 90(4) (2006).

Jon R. Katzenbach and Douglas K. Smith, *The Discipline of Teams: A Mindbook-Workbook for Delivering Small Group Performance* (New York: Wiley, 2001).

James R. Larson Jr., *In Search of Synergy in Small Group Performance* (London: Psychology Press, 2009).

Charles Day, "Right-Sizing the Table," *Association Management*, July 28, 2005.

Earl A. Alluisi, *The Measurement of Small Group Performance in a Restrictive Environment* (Lockheed, GA: Human Factors Research Department, 1962).

Francis L. Ulschak, *Small Group Problem Solving: An Aid to Organizational Effectiveness* (Reading, MA: Addison-Wesley, 1981).

Nancy K. Napier, *Insight: Encouraging Aha! Moments for Organizational Success* (Westport, CT: Greenwood, 2010).

p. 246 Between 1995 and 2004, Johnson & Johnson acquired 51 smaller companies
Stephanie Saul, Market Place: Buyer's Remorse Is Causing Some Palpitations at Johnson," *New York Times*, November 3, 2005.

Arlene Weintraub, "Johnson and Johnson's Next Baby?" *BusinessWeek*, June 18, 2007.

Neal Chatigny, Kevin Higginbotham, John Walsh, and Kyle Williams, "Growth Strategies in the Pharmaceutical Industry: Strategic Acquisitions," April 29, 2003, www. mcafee.cc/Classes/BEM106/Papers/UTexas/2003/JandJ.pdf.

Matthew Herper, "For Drug Deals, Think Small," *Forbes*, February 10, 2003.

p. 246 Dot com legend Google is on the same path
"List of Acquisitions by Google" (SEO Consultants, July 3, 2007).

Jess Chan, "Google's Acquisition Strategy," fishtrain.com, September 13, 2007, http:// fishtrain.com/2007/09/13/googles-acquisition-strategy/.

Larry Dignan, Sam Diaz, and Tom Steinert-Threlkeld, "Google's Acquisition Strategy Should Think Small and Mobile," Blogs.zdnwt.com, April 12, 2007, www.zdnet. com/blog/btl/googles-acquisition-strategy-should-think-small-and-mobile/4827.

Randall E. Stross, *Planet Google: One Company's Audacious Plan to Organize Everything We Know* (New York: Free Press, 2008).

p. 246 The renowned Tennessee Student/Teacher Achievement Ratio
Elizabeth Word, Helen Pate Bain, B. DeWayne Fulton, Jane Boyd Zaharias, Charles M. Achilles, Martha Nannette Lintz, John Folger, and Carolyn Breda, "The State of Tennessee's Student/Teacher Achievement Ratio (STAR) Project: Final Summary, 1985–1990," www.misd.k12.wa.us/departments/superintendent/class_size/documents/ STARReport.pdf.

Samuel L. Blumenfeld, *The Whole Language/OBE Fraud: The Shocking Story of How America Is Being Dumbed Down by Its Own Education System* (Boulder, CO: Paradigm Publishers, 1995).

Robert J. Manley, *Designing School Systems for All Students: A Tool Box to Fix America's Schools* (Lanham, MD: Rowman and Littlefield Education, 2009).

Hank Kraychir and Jim Garvey, *Destruction of the Education Monster: The Destruction of America's Public School System* (CreateSpace, 2008).

Richard Rothstein, Lawrence Misitel, Jennifer King Rice, and Eric A. Hanushek, "The Class Size Debate" (Washington, DC: Economic Policy Institute, June 27, 2002).

John W. Alspaugh, "The Relationship Between School Size, Student-Teacher Ratio, and School Efficiency" (Project Innovation, July 28, 2005).

J. D. Finn and C. M. Achilles, "Answers and Questions About Class Size: A Statewide Experiment," *American Educational Research Journal*, Fall 1990.

p. 247 Teaching a robot to walk on bumpy terrain like cobblestones is so challenging
"Robot Unravels Mystery of Walking," BBC News, July 12, 2007.

Kimberly Patch, "Humanoid Robots Walk Naturally," *Technology Research News*, February 23, 2005.

"Walk This Way: The Amazing Complexity of Getting Around," July 16, 2007, www.livescience.com/animals/070716_walking_sidebar.html.

Mark Randall, *Adaptive Neural Control of Walking Robots* (New York: Wiley, 2001).

Craig Stanford, *Upright: The Evolutionary Key to Becoming Human* (New York: Houghton Mifflin, 2003).

Enric Celaya and Joseph M. Porta, "A Control Structure for the Locomotion of a Legged Robot on Difficult Terrain" (Barcelona, Spain: Institut de Robòtica i Informàtica Industrial [*IEEE Robotics and Automation Magazine* 5(2) (1998)]).

Takashi Takuma and Koh Hosoda, "Terrain Negotiation of a Compliant Biped Robot Driven by Antagonistic Artificial Muscles," *Journal of Robotics and Mechatronics* 19(4) (2007).

p. 248 Walking on *any* surface has cognitive benefits

Arthur Kramer, "Healthy Body, Healthy Mind? The Relationship Among Fitness, Cognition, Brain Structure, and Function" (National Institute on Aging Symposium on Cognitive Training for Older Adults, Bethesda, MD, February 29, 2004).

Michael Logan, "A User's Guide to Life-Long Brain Health and Fitness," askmikethe counselor2.com, http://brainfitness.boomja.com/index.php?ITEM=59007.

Oregon Research Institute, comp., "Oregon Study Confirms Health Benefits of Cobblestone Walking for Older Adults," *Science Daily*, June 30, 2005.

Mary Carmichael, "Exercise Does More Than Just Build Muscles," *Newsweek*, March 19, 2007.

National Institutes of Health, "Lessons Learned from Couch Mice, Marathon Mice, and Men and Women Who Like to Walk," November 25, 2008, www.nia.nih.gov/Alzheimers/Publications/ADProgress2005_2006/Part2/lessons.htm.

N. Jausovec and K. Bakracevic, *Creativity Research Journal* 8(1) (January 1995).

p. 249 The human brain operates best when it is regularly subjected to new challenges

Jane Brody, "Mental Reserves Keep Brains Agile," *New York Times*, December 11, 2007.

"Learning New Things Vital to Brain Health," *Sydney Morning Herald*, March 18, 2009.

"Use It or Lose It? Study Suggests Mentally Stimulating Activities May Reduce Alzheimer's Risk," February 13, 2002, www.prohealth.com/library/showarticle.cfm?libid=4475.

Joene Hendry, "Keeping the Aging Mind Active Cuts Dementia Risk," Reuters, September 23, 2009.

"Doing Crosswords Puzzles May Help Delay Alzheimer's Onset" (New York: Fisher Center for Alzheimer's Research Foundation, June 5, 2010).

"14 Scientifically (Research-)Proven Ways to Boost Brain Power," emedexpert.com.

"83 Unique Exercises to Keep Your Brain Alive," emedexpert.com.

Caroline Latham, "Improve Brain Health Now: Easy Steps," SharpBrains, April 11, 2009, www.sharpbrains.com/blog/2007/04/11/easy-steps-to-improve-your-brain-health-now/.

Michael Logan, "A User's Guide to Life-Long Brain Health and Fitness," askmikethe counselor2.com, http://brainfitness.boomja.com/index.php?ITEM=59007.

John Ratey and Eric Hagerman, *Spark: The Revolutionary New Science of Exercise and the Brain* (New York: Little, Brown, 2008).

Frawley Bridwell, "Study Shows Link Between Morbid Obesity, Low IQ in Toddlers" (Gainesville: College of Medicine, University of Florida, August 2008).

Kathleen Phalen Tomaselli, "Steps to a Nimble Mind: Physical and Mental Exercise Help Keep the Brain Fit," amednews.com, November 11, 2008, www.ama-assn.org/amednews/2008/11/17/hlsa1117.htm.

p. 250 Mike Logan, an education counselor at Illinois State University, offers some easy suggestions

Michael Logan, "A User's Guide to Life-Long Brain Health and Fitness," askmikethe counselor2.com, http://brainfitness.boomja.com/index.php?ITEM=59007.

"83 Unique Exercises to Keep Your Brain Alive," emedexpert.com.

Kathleen Phalen Tomaselli, "Steps to a Nimble Mind: Physical and Mental Exercise Help Keep the Brain Fit," amednews.com, November 11, 2008, www.ama-assn.org/amednews/2008/11/17/hlsa1117.htm.

Author interviews with retired postal workers who originally participated in the program.

p. 252 The Japanese government gets it
Mitsuo Kawato, "Understanding the Brain by Creating the Brain: Toward Manipulative Neuroscience," http://rstb.royalsocietypublishing.org/content/363/1500/2201.abstract.
Grace Wong, "Boom Times for Brain Training Games," CNN, December 11, 2008.
"Ultimate Boomer Self-Employment: Becoming a Brain Trainer via PositScience," April 17, 2008, Boomers.typepad/2008.
Alvaro Fernandez, "Brain Fitness and Exercise in Japan," SharpBrains, March 21, 2007, www.sharpbrains.com/blog/2007/03/21/brain-fitness-and-exercise-in-japan/.
Dan Choi, "Japanese Doctors Recommend Brain Training for Seniors," Joystiq, March 7, 2006, www.joystiq.com/2006/03/07/japanese-doctors-recommend-brain-training-for-seniors/.
Go Hirano, "Brain Training and Mind Games: Interview with Japanese Expert," SharpBrains, December 8, 2007.
Sean McCall, "English + Brain Science = A Japanese Clockwork Orange," *ESL Daily*, March 27, 2008.
Masao Ito, "Japanese Science Funding," *Science*, May 16, 1997.
"Basic Research and Science in the Japanese Economy," ww.jei.org/AJAclass/Science R&D.pdf.
Hiromi Mizuno, *Science for the Empire: Scientific Nationalism in Modern Japan* (Palo Alto, CA: Stanford University Press, 2008).
p. 253 Playtime had a dramatic effect on intelligence tests
E. Saltz, D. Dixon, and J. Johnson, "Training disadvantaged preschoolers on various fantasy activities: Effects on cognitive functioning and impulse control," *Child Development* 48(20) (1977): 367–380. (ERIC Journal No. EJ164702).
Susan Ohanian, *What Happened to Recess and Why Are Our Children Struggling in Kindergarten?* (New York: McGraw-Hill, 2007).
John Ratey and Eric Hagerman, *Spark: The Revolutionary New Science of Exercise and the Brain* (New York: Little, Brown, 2008).
Frawley Bridwell, "Study Shows Link Between Morbid Obesity, Low IQ in Toddlers" (Gainesville: College of Medicine, University of Florida, August 2008).
p. 253 The positive effect recess has on learning
Anthony D. Pellegrini and Robyn M. Holmes,"The Role of Recess in Primary School," May 16, 2005, http://udel.edu/~roberta/play/Pellegrini.pdf, May 16, 2005.
p. 254 Stories such as the spilled can of paint
Matthew May, "Breakthrough by Taking Breaks," *American Express Open Forum*, September 17, 2009.
"83 Unique Exercises to Keep Your Brain Alive," emedexpert.com.
p. 254 "The relaxation phase is crucial"
Kalina Christoff, Alan Gordon, and Rachelle Smith, "The Role of Spontaneous Thought in Human Cognition," in *Neuroscience of Decision Making*, ed. Oshin Vartanian and David R. Mandel (New York: Psychology Press, n.d.).
Karuna Subramaniam, John Kounios, B. Todd, and Mark Jung-Beeman, "A Brain Mechanism for Facilitation of Insight by Positive Affect," *Journal of Cognitive Neuroscience* 21(3) (2009).
B. R. Sheth, S. Sandkühler, J. Bhattacharya, "Posterior Beta and anterior gamma oscillations predict cognitive insight." *Journal of Cognitive Neuroscience*, 21(7) (July 2009): 1269–1279.
John Kounios, Jessica I. Fleck, Deborah L. Green, Lisa Payne, Jennifer L. Stevenson, Ed-

ward M. Bowden, and Mark Jung-Beeman, "The Origins of Insight in Resting-State Brain Activity" (Philadelphia: Department of Psychology, Drexel University, July 18, 2007).

J. S. Antrobus, J. L. Singer, and S. Greenberg, "Studies in the Stream of Consciousness: Experimental Enhancement and Suppression of Spontaneous Cognitive Processes," *Perceptual and Motor Skills* 23 (1966).

Interview with Jerry Lauch, Louisville Sleep Disorders Center, Carmel, CA, 2009.

A. Dijksterhuis and T. Meurs, "Where Creativity Resides: The Generative Powers of Unconscious Thought," *Consciousness and Cognition* 15(1) (March 2006).

Howard Shevrin, James A. Bond, Linda A. Grakel, Richard K. Hertel, and William Williams, *Conscious and Unconscious Processes: Psychodynamic, Cognitive, and Neurophysiological Convergences* (New York: Guilford Press, 1996).

Dan J. Stein, *Cognitive Science and the Unconscious* (Washington, DC: American Psychiatric Press, 1997).

Gerd Gigerenzer, *Gut Feelings: The Intelligence of the Unconscious*, narr. Dick Hill (Old Saybrook, CT: Tantor Media Audio Books, 2007).

Geoffrey Underwood, ed., *Implicit Cognition* (New York: Oxford University Press, 1996).

J. R. Binder, J. A. Frost, T. A. Hammeke, P. S. Bellgowan, S. Rao, and Mand Cox, "Conceptual Processing During the Conscious Resting State: A Functional MRI Study," *Journal of Cognitive Science* 11(9) (1999).

p. 254 One of the predictors of insightful thinking is the appearance of alpha waves
Lee Holtz, "A Wandering Mind Heads Straight Toward Insight," *Wall Street Journal*, June 19, 2009.

p. 255 Research has demonstrated that the brain "prepares" itself
Karuna Subramaniam, John Kounios, B. Todd, and Mark Jung-Beeman, "A Brain Mechanism for Facilitation of Insight by Positive Affect," *Journal of Cognitive Neuroscience* 21(3) (2009).

J. Kounios, J. L. Frymiare, E. M. Bowden, J. I. Fleck, K. Subramaniam, T. B. Parnish, and M. Jung-Beeman, "The Prepared Mind: Neural Activity Prior to Problem Presentation Predicts Subsequent Solution by Sudden Insight," *Psychological Sciences* 17(10) (2006).

A. J. K. Pols, "Insight in Problem Solving," http://www.phil.uu.nl/preprints/ckiscripties/SCRIPTIES/018_pols.pdf.

Kalina Christoff, Alan Gordon, and Rachelle Smith, "The Role of Spontaneous Thought in Human Cognition," in *Neuroscience of Decision Making*, ed. Oshin Vartanian and David R. Mandel (New York: Psychology Press, n.d.).

John Kounios, Jessica I. Fleck, Deborah L. Green, Lisa Payne, Jennifer L. Stevenson, Edward M. Bowden, and Mark Jung-Beeman, "The Origins of Insight in Resting-State Brain Activity" (Philadelphia: Department of Psychology, Drexel University, July 18, 2007).

William James, *On Vital Reserves: The Energies of Men, the Gospel of Relaxation* (New York: Henry Holt, 1922).

Jerry Swartz, "The Conscious 'Pop': A Nonconscious Processing Framework for Problem Solving" (Cold Spring Harbor, NY: Cold Spring Harbor Laboratory, New Frontiers in Studies of the Unconscious, April 9, 2007).

p. 258 So, the first priority for getting the highest brain performance
"Top 5 Brain Health Foods," BrainReady, CNN, May 7, 2009.

David Zinczenko and Matt Goulding, "Best and Worst Brain Foods," *Men's Health*, November 1, 2009.

"Age and the Brain—Nature Brain Foods for Memory: Best Brain Food, Good Food for the Brain," www.add-adhd-helpcenter.com.

Brain Food (The Thinking Business, 2009).

"Weekly Curry May Fight Dementia," BBC, June 3, 2009.

"14 Scientifically Proven Ways to Boost Brain Power," *Mindpower News*, March 4, 2010, www.mindpowernews.com/ProvenBrainPower.htm.

Lorraine Perretta and Oona Van den Berg, *Brain Food: The Essential Guide to Boosting Brain Power* (New York: Sterling, 2002).

"83 Unique Exercises to Keep Your Brain Alive," emedexpert.com.

p. 258 The impact of physical health plays a large role in the brain's readiness

Arthur F. Kramer, Kirk I. Erickson, and Stanley J. Colcombe, "Exercise, Cognition, and the Aging Brain," *Journal of Applied Psychology*, June 15, 2006.

John Ratey, presentation at Renaissance Conference, Monterey, CA, September 4, 2009.

Interview with John Ratey, Cambridge, MA, 2009.

Terry McMorris, Philli Tomporowski, and Michel Audiffren, *Exercise and Cognitive Function* (Hoboken, NJ: Wiley, 2009).

Waneen W. Spirduso, Leonard W. Poon, and Wojtek Chodzko-Zajko, eds., *Exercise and Its Mediating Effects on Cognition* (Champaign, IL: Human Kinetics, 2008).

John Ratey and Eric Hagerman, *Spark: The Revolutionary New Science of Exercise and the Brain* (New York: Little, Brown, 2008).

Frawley Bridwell, "Study Shows Link Between Morbid Obesity, Low IQ in Toddlers" (Gainesville: College of Medicine, University of Florida, August 2008).

Arthur Kramer, "Healthy Body, Healthy Mind? The Relationship Among Fitness, Cognition, Brain Structure, and Function" (National Institute on Aging Symposium on Cognitive Training for Older Adults, Bethesda, MD, February 29, 2004).

Michael Logan, "A User's Guide to Life-Long Brain Health and Fitness," askmikethecounselor2.com, http://brainfitness.boomja.com/index.php?ITEM=59007.

Mary Carmichael, "Exercise Does More Than Just Build Muscles," *Newsweek*, March 19, 2007.

"14 Scientifically Proven Ways to Boost Brain Power," *Mindpower News*, March 4, 2010, www.mindpowernews.com/ProvenBrainPower.htm.

Kathleen Phalen Tomaselli, "Steps to a Nimble Mind: Physical and Mental Exercise Help Keep the Brain Fit," amednews.com, November 11, 2008, www.ama-assn.org/amednews/2008/11/17/hlsa1117.htm.

D. S. Albeck, K. Sano, G. E. Prewitt, and L. Dalton, *Behavior Brain Research*, 2006.

S. J. Colcombe, A. F. Kramer, K. I. Erickson, P. Scalf, E. McAuley, N. J. Cohen, A. Webb, G. J. Jerome, D. X. Marquez, and S. Elavsky, *Proceedings of National Academy of Sciences*, 2004.

C. Fabre, K. Charmi, P. Mucci, J. Masse-Biron, and C. Prefaut, *Internal Sports Medicine*, 2002.

J. Farmer, X. Zhao, H. Van Praag, K. Wodtke, F. H. Gage, and B. R. Christie, *Neuroscience*, 2004.

Y. P. Kim, H. Kim, M.S. Shin, H. K. Chang, M. H. Jang, M. C. Shin, S. J. Lee, H. H. Lee, J. H. Yoon, I. G. Jeong, and C. J. Kim, *Neuroscience Letters*, 2004.

A. F. Kramer, S. J. Colcombe, K. I. Erickson, and P. Paige, "Fitness Training and the

Brain: From Molecules to Minds" (proceedings of the 2006 Cognitive Aging Conference, Georgia Institute of Technology, Atlanta, GA, 2006).

K. Yaffe, A. J. Fiocco, K. Lundquist, E. Vittinghoff, E. M. Simonsick, A. B. Newman, S. Satterfield, C. Rosano, S. M. Rubin, H .N. Ayonayon et al., "Predictors of Maintaining Cognitive Function in Older Adults: The Health ABC Study," *Neurology*, June 9, 2009.

J. D. Williamson, M. Espeland, S. B. Kritchevsky, A. B. Newman, A. C. King, M. Pahor, J. M. Guralnik, L. A. Pruitt, and M. E. Miller, "Changes in Cognitive Function in a Randomized Trial of Physical Activity: Results of the Lifestyle Interventions and Independence for Elders Pilot Study," *Journals of Gerontology, Series A*, 2009.

K. I. Erickson and A. F. Kramer, "Aerobic Exercise Effects on Cognition and Neural Plasticity in Older Adults," *Sports Medicine*, January 1, 2009.

p. 260 The harm that the absence of variety causes

Presentation by Michael Merzenich at Oregon State University, Corvallis, May 1, 2009.

Interview with Dr. Michael Merzenich, University of California Medical Center, Keck Center, San Francisco, 2009.

Michael Merzenich, "Does Exercise Make Kids Smarter?" PositScience, March 30, 2007, http://merzenich.positscience.com/2007/03/30/march-30-does-exercise-make-kids-smarter/.

p. 260 75 percent of Americans have trouble sleeping

Anastacia Mott Austin, "Americans Just Not Getting Enough Sleep, Study Shows," Buzzle.com, March 6, 2008, www.buzzle.com.

Centers for Disease Control, "Insufficient Rest or Sleep in Adults, United States, 2008," www.cdc.gov/mmwr/preview/mmwrhtml/mm5842a2.htm.

U. Wagner, S. Gais, H. Haider, R. Verleger, and J. Born, "Sleep Inspires Insight," *Nature* 427 (January 22, 2004).

J. M. Ellenbogen, P. T. Hu, J. D. Payne, Titone, and M. P. Walker, "Human Relational Memory Requires Time and Sleep," *Proceedings of the National Academy of Sciences* 104(18) (May 1, 2007).

S. Banks and D. F. Dinges, "Behavioral and Physiological Consequences of Sleep Restriction," *Journal of Clinical Sleep Medicine* 3(5) (2007).

Harvey R. Colten and Bruce M. Altevogt, eds., *Sleep Disorders and Sleep Deprivation: An Unmet Public Health Problem* (Washington, DC: National Academies Press, 2006).

"Got a Problem? Think About It Overnight," *US News and World Report*, June 9, 2009.

Jerry Swartz, "The Conscious 'Pop': A Nonconscious Processing Framework for Problem Solving" (Cold Spring Harbor, NY: Cold Spring Harbor Laboratory, New Frontiers in Studies of the Unconscious, April 9, 2007).

Richard R. Bootzin, John F. Kihlstrom, and Daniel L. Schacter, eds., *Sleep and Cognition* (Washington, DC: American Psychological Association, January 1994).

Interview with Jerry Lauch, Louisville Sleep Disorders Center, Carmel, CA, 2009.

A. Dijksterhuis and T. Meurs, "Where Creativity Resides: The Generative Powers of Unconscious Thought," *Consciousness and Cognition* 15(1) (March 2006).

Howard Shevrin, James A. Bond, Linda A. Grakel, Richard K. Hertel, and William Williams, *Conscious and Unconscious Processes: Psychodynamic, Cognitive, and Neurophysiological Convergences* (New York: Guilford Press, 1996).

Dan J. Stein, *Cognitive Science and the Unconscious* (Washington, DC: American Psychiatric Press, 1997).

Gerd Gigerenzer, *Gut Feelings: The Intelligence of the Unconscious*, narr. Dick Hill (Old Saybrook, CT: Tantor Media Audio Books, 2007).

Geoffrey Underwood, ed., *Implicit Cognition* (New York: Oxford University Press, 1996).

"83 Unique Exercises to Keep Your Brain Alive," emedexpert.com.

H. Fiss, E. Kremer, and J. Litchman, The Mnemonic Function of Dreaming," *Sleep Research* 6(122) (1977).

M. J. Fosse, R. Fosse, J. A. Hobson, and R. J. Stickgold, "Dreaming and Episodic Memory: A Dysfunctional Dissociation?" *Journal of Cognitive Neuroscience*, 2003.

Ernest Hartman, *Dreams and Nightmares: The New Theory on the Origin and Meaning of Dreams* (New York: Plenum Press, 1998).

P. Maquet, "The Role of Sleep in Learning and Memory," *Science*, 2001.

p. 261 Antidepressants are now the most prescribed medication

Elizabeth Cohen, "CDC: Antidepressants Most Prescribed Drugs in United States," CNN, 2008.

Shankar Vedantam, "Antidepressants Use by U.S. Adults Soars," *Washington Post*, December 3, 2004.

"Drug and Alcohol Rehab Industry Grows to 7.7 Billion This Year, with It 1+ Million Americans in Treatment," Marketdata, October 30, 2000.

Bobbie Hasselbring, "How Do Antidepressants Work?" Health.discovery.com, May 5, 2009, http://health.discovery.com/centers/articles/articles.html?chrome=c09&article=LC_33&center=p06.

Jerome Yesavage, "Memory Training in Older Adults: Issues of Prediction of Response, Mild Impairment, and Interaction with Medications" (National Institute of Aging Symposium on Cognitive Training for Older Adults, Bethesda, MD, February 29, 2004).

Richard Bentall, *Doctoring the Mind: Is Our Treatment of Mental Illness Really Any Good?* (New York: New York University Press, 2009).

Dawson Hodges and Colin Burchfield, *Mind, Brain, and Drugs: An Introduction to Psychopharmacology* (Boston: Allyn and Bacon, 2005).

Joanna Moncrieff, *The Myth of a Chemical Cure: A Critique of Psychiatric Drug Treatment* (New York: Palgave Macmillan, 2009).

Ellliot Valenstein, *Blaming the Brain: The Truth About Drugs and Mental Health* (New York: Free Press, 2002.

Chapter 13.
On the Threshold

p. 268 Here, visionary Paul Hawken gets the final word

Paul Hawken, "The Class of 2009 Commencement Address," University of Portland, Portland, OR, May 3, 2009, www.charityfocus.org.

Index